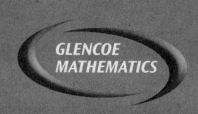

GLENCOE
MATHEMATICS

# Skills Inte...
## for Pre-Algebra
### Diagnosis and Remediation

MW00699807

# Student Workbook

McGraw Hill **Glencoe**

New York, New York    Columbus, Ohio    Chicago, Illinois    Peoria, Illinois    Woodland Hills, California

**Mc Graw Hill** **Glencoe**

*The McGraw·Hill Companies*

Send all inquiries to:
Glencoe/McGraw-Hill
8787 Orion Place
Columbus, OH 43240-4027

ISBN: 0-07-867808-0

*Pre-Algebra Intervention*
*Student Workbook*

3 4 5 6 7 8 9 10   009   11 10 09 08 07 06

# Table of Contents

**Skill**    **Number and Operation**

| | | |
|---|---|---|
| 1 | Place Value | 1 |
| 2 | Rounding Numbers | 3 |
| 3 | Order of Operations | 5 |
| 4 | Properties | 7 |
| 5 | Integers | 9 |
| 6 | Adding and Subtracting Integers | 11 |
| 7 | Multiplying and Dividing Integers | 13 |
| 8 | Prime Factorization | 15 |
| 9 | Divisibility Patterns | 17 |
| 10 | Greatest Common Factor | 19 |
| 11 | Least Common Multiple | 21 |
| 12 | Adding and Subtracting Decimals | 23 |
| 13 | Multiplying and Dividing Decimals | 25 |
| 14 | Adding and Subtracting Fractions | 27 |
| 15 | Multiplying and Dividing Fractions | 29 |
| 16 | Changing Fractions to Decimals | 31 |
| 17 | Percents as Fractions and Decimals | 33 |
| 18 | Percent of a Number | 35 |
| 19 | Percent Proportion | 37 |
| 20 | Percent of Change | 39 |
| 21 | Powers and Exponents | 41 |
| 22 | Scientific Notation | 43 |

**Skill**    **Algebra**

| | | |
|---|---|---|
| 23 | Variables and Expressions | 45 |
| 24 | Writing Expressions and Equations | 47 |
| 25 | Simplifying Expressions and Equations | 49 |
| 26 | Solve Equations Involving Addition and Subtraction | 51 |
| 27 | Solve Equations Involving Multiplication and Division | 53 |
| 28 | Solve Two-Step Equations | 55 |
| 29 | Solve Inequalities | 57 |
| 30 | Ratio and Proportion | 59 |
| 31 | Proportional Reasoning | 61 |
| 32 | Scale Drawings | 63 |
| 33 | Ordered Pairs and the Coordinate Plane | 65 |
| 34 | Function Tables | 67 |
| 35 | Graphing Functions | 69 |
| 36 | Ratios and Rates | 71 |
| 37 | Slope of a Line | 73 |
| 38 | Graphing Linear Equations | 75 |
| 39 | Solve Equations in Two Variables | 77 |
| 40 | Square Roots | 79 |
| 41 | Sequences | 81 |

| Skill | Geometry | |
|---|---|---|
| 42 | Geometric Terms | .83 |
| 43 | Angles | .85 |
| 44 | Angle Relationships | .87 |
| 45 | Parallel Lines and Angle Relationships | .89 |
| 46 | Triangles and Quadrilaterals | .91 |
| 47 | Similar Figures | .93 |
| 48 | Congruent Figures | .95 |
| 49 | Translations | .97 |
| 50 | Reflections | .99 |
| 51 | The Pythagorean Theorem | 101 |
| 52 | Three-Dimensional Figures | 103 |

| Skill | Measurement | |
|---|---|---|
| 53 | Customary Units of Measure | 105 |
| 54 | Metric Units of Measure | 107 |
| 55 | Converting Between Customary and Metric Units | 109 |
| 56 | Units of Time | 111 |
| 57 | Precision and Significant Digits | 113 |
| 58 | Perimeter and Area | 115 |
| 59 | Area of Parallelograms, Rectangles, and Squares | 117 |
| 60 | Area of Triangles and Trapezoids | 119 |
| 61 | Area of Irregular Shapes | 121 |
| 62 | Circumference and Area of Circles | 123 |
| 63 | Nets and Solids | 125 |
| 64 | Volume of Rectangular Prisms and Cylinders | 127 |
| 65 | Surface Area of Rectangular Prisms and Cylinders | 129 |

| Skill | Data Analysis and Probability | |
|---|---|---|
| 66 | Mean, Median, and Mode | 131 |
| 67 | Frequency Tables | 133 |
| 68 | Line Plots | 135 |
| 69 | Stem-and-Leaf Plots | 137 |
| 70 | Box-and-Whisker Plots | 139 |
| 71 | Line Graphs | 141 |
| 72 | Bar Graphs and Histograms | 143 |
| 73 | Circle Graphs | 145 |
| 74 | Using Statistics to Make Predictions | 147 |
| 75 | Counting Outcomes and Tree Diagrams | 149 |
| 76 | Permutations | 151 |
| 77 | Combinations | 153 |
| 78 | Probability | 155 |
| 79 | Theoretical and Experimental Probability | 157 |

| Skill | Problem Solving | |
|---|---|---|
| 80 | Problem-Solving Strategies | 159 |
| 81 | Work Backward | 161 |
| 82 | Simplify and Use Logical Reasoning | 163 |
| 83 | Organizing Information | 165 |
| 84 | Visualizing Information | 167 |

**Name** _____ **Date** _____

# Place Value

$E$ach digit of a number represents a specific value.

**EXAMPLE** *In 2004, the estimated world population was 6,342,000,000 people. Write this number in words and in expanded notation.*

| Billions | | | Millions | | | Thousands | | | Ones | | |
|---|---|---|---|---|---|---|---|---|---|---|---|
| hundreds | tens | ones | hundreds | tens | ones | hundreds | tens | ones | hundreds | tens | ones |
| | | 6, | 3 | 4 | 2, | 0 | 0 | 0, | 0 | 0 | 0 |

The number 6,342,000,000 may be written in these ways.
**Standard Form:**       6,342,000,000
**Words:**                 six billion, three hundred forty-two million
**Expanded Notation:** 6,000,000,000 + 342,000,000

**EXERCISES**   *Write the number named by each underlined digit.*

1. <u>2</u>4

2. 92<u>7</u>

3. <u>3</u>,962

4. 31,<u>2</u>47

5. 124,8<u>5</u>7

6. 1<u>6</u>7,000

7. <u>8</u>,240,000

8. 1,<u>4</u>32,512,454

*Write each number in words.*

9. 387

10. 502

11. 8,043

12. 10,350

13. 420,000

14. 608,000

15. 2,661,075

16. 713,004,112

17. 12,100,562,008

*Write each number in expanded notation.*

**18.** 287

**19.** 820

**20.** 4,130

**21.** 79,321

**22.** 604,503

**23.** 10,352,718

**24.** 1,098,672,442

**25.** 542,098,404,000

*Fill in each blank with <, >, or = to make a true sentence.*

**26.** 478,204 _____ 478,195          **27.** 547,932 _____ 548,000

**28.** 5,000,000 _____ 500,000          **29.** 3,890,000,000 _____ 400,000,000

**30.** 579,872,221 _____ 47,987,222,011          **31.** 53,872,014,454 _____ 53,872,014,454

**32.** 418,000,000 _____ 420,000,000          **33.** 700,612,000,090 _____ 699,999,999,999

## APPLICATIONS

**34.** The distance from Jacksonville, Florida, to Pittsburgh, Pennsylvania, is 865 miles. Write 865 in expanded form.

**35.** The Mississippi River is 2,348 miles long. Write this distance in words.

*Use the table at the right.*

**36.** Which ocean has the greater area, the Indian or the Arctic Ocean?

**37.** Rank the oceans in order of size. The ocean with the greatest area gets a rank of 1.

**38.** Which oceans have areas less than 76,000,000 square kilometers?

| Areas of Oceans | |
|---|---|
| **Ocean** | **Area (sq km)** |
| Arctic | 14,056,000 |
| Atlantic | 76,762,000 |
| Indian | 68,556,000 |
| Pacific | 155,557,000 |
| Southern | 20,327,000 |

**Name** _____    **Date** _____

# Order of Operations

**W**hen you evaluate an expression in mathematics, you must perform the operations in a certain order. This order is called the **order of operations**.

| Order of Operations |
|---|
| **1.** Do all operations within grouping symbols first; start with the innermost grouping and work out. |
| **2.** Evaluate all powers before other operations. |
| **3.** Multiply and divide in order from left to right. |
| **4.** Add and subtract in order from left to right. |

**EXAMPLE**    *Evaluate $8 + 56 \div [(5 - 3) \times 4] - 12$.*

$$8 + 56 \div [(5 - 3) \times 4] - 12 = 8 + 56 \div [2 \times 4] - 12 \quad \text{Subtract 3 from 5 inside the parentheses.}$$

$$= 8 + 56 \div 8 - 12 \quad \text{Multiply 2 and 4.}$$

$$= 8 + 7 - 12 \quad \text{Divide 56 by 8.}$$

$$= 15 - 12 \quad \text{Add 8 and 7.}$$

$$= 3 \quad \text{Subtract 12 from 15.}$$

**EXERCISES**    *Name the operation that should be done first in each expression.*

**1.** $(9 + 3) \times 7$     **2.** $98 - 5 \times 7$     **3.** $5 \times (9 - 1)$

**4.** $(15 \div 3) + (4 + 5)$     **5.** $5 \times 4 \div 2$     **6.** $5(5 - 3) \times 2$

**7.** $5 + 4 \cdot 7$     **8.** $13(6 + 3)$     **9.** $(4 - 2) + 6$

**10.** $6 \times 8 \div 4$     **11.** $32 \div 4 \times 2$     **12.** $9(4 + 2) \div 3$

*Evaluate each expression.*

**13.** $8 \cdot 7 + 8 \cdot 3$

**14.** $(9 - 3) \div 3$

**15.** $8 - 6 + 3$

**16.** $18 \div 3 \cdot 6$

**17.** $24 \div (6 - 2)$

**18.** $9 - 4 \div 2 + 6$

**19.** $9 \cdot 3 + 8 \div 4$

**20.** $3(18 - 12) - (5 - 3)$

**21.** $(32 + 8) \times 3 \div 4$

**22.** $(15 - 3) \div (1 \cdot 6)$

**23.** $18 \div (3 + 3) - 2$

**24.** $(16 - 8) \div 4 + 10$

**25.** $5 \cdot (16 \div 4) - 4 \div 2$

**26.** $56 \div [4 \times (5 - 3)]$

**27.** $6 \times [1 + (5 - 2)]$

**28.** $15 - 2[4 - (6 - 3)]$

**APPLICATIONS** *Use the prices to write a mathematical expression for each total cost. Evaluate the expression to find the total cost.*

**29.** 3 sandwiches, and 3 side items

**30.** 4 sandwiches, 2 side items, and 1 bag of chips

| Ridgeview Deli Menu | |
| --- | --- |
| Sandwich | $4.50 |
| Side Item | $2.25 |
| Chips | $0.75 |
| Dessert | $2.50 |

**31.** 4 sandwiches and 4 desserts with a coupon for $1.00 off each sandwich

**32.** 2 sandwiches, 2 side items, 1 bag of chips, and 2 desserts with a coupon for $2.00 off any total

**33.** 6 sandwiches, 3 side items, and 5 desserts with a coupon for a free sandwich for every one that is purchased

Name _____ Date _____

# Properties

The table shows the properties for addition and multiplication.

| Property | Examples | |
|---|---|---|
| **Commutative**<br>The sum or product of two numbers is the same regardless of the order in which they are added or multiplied. | **Addition**<br>$2 + 3 = 3 + 2$<br>$5 = 5$ | **Multiplication**<br>$4 \times 6 = 6 \times 4$<br>$24 = 24$ |
| **Associative**<br>The sum or product of three or more numbers is the same regardless of the way in which they are grouped. | **Addition**<br>$(5 + 2) + 6 = 5 + (2 + 6)$<br>$7 + 6 = 5 + 8$<br>$13 = 13$ | **Multiplication**<br>$(3 \cdot 4) \cdot 7 = 3 \cdot (4 \cdot 7)$<br>$12 \cdot 7 = 3 \cdot 28$<br>$84 = 84$ |
| **Distributive**<br>The sum of two addends multiplied by a number is equal to the products of each addend and the number. | $5 \cdot (6 + 2) = (5 \cdot 6) + (5 \cdot 2)$<br>$5 \cdot (8) = 30 + 10$<br>$40 = 40$ | |
| **Identity Property of Addition**<br>The sum of a number and 0 is the number. | $9 + 0 = 9$ | |
| **Identity Property of Multiplication**<br>The product of a number and 1 is the number. | $15 \times 1 = 15$ | |
| **Inverse Property of Addition**<br>The sum of a number and its additive inverse (opposite) is 0. | $4 + (-4) = 0$ | |
| **Inverse Property of Multiplication**<br>The product of a number and its multiplicative inverse (reciprocal) is 1. | $2 \times \frac{1}{2} = \frac{2}{1} \times \frac{1}{2}$<br>$= 1$ | |

**EXERCISES** *Name the additive inverse, or opposite of each number.*

**1.** 8      **2.** 5      **3.** $\frac{3}{4}$      **4.** $1\frac{1}{2}$

*Name the multiplicative inverse, or reciprocal of each number.*

**5.** 4      **6.** 7      **7.** $\frac{2}{5}$      **8.** $\frac{7}{16}$

**Name the property shown by each statement.**

**9.** $34 + 42 = 42 + 34$

**10.** $8 \times (53 + 12) = (8 \times 53) + (8 \times 12)$

**11.** $\frac{1}{16} \times 16 = 1$

**12.** $16 \cdot (5 \cdot 15) = (16 \cdot 5) \cdot 15$

**13.** $\frac{2}{5} \cdot \frac{5}{3} = \frac{5}{3} \cdot \frac{2}{5}$

**14.** $(32 + 48) + 52 = 32 + (48 + 52)$

**15.** $256 + 0 = 256$

**16.** $\frac{3}{10} \cdot \frac{10}{3} = 1$

**17.** $1 \times 143 = 143$

**18.** $81 + (-81) = 0$

## APPLICATIONS

**19.** Michael rides his bike $2\frac{3}{5}$ as long as Jacob. Find Michael's riding time if Jacob rides for 45 minutes.

**20.** A daisy is 24 inches tall. The height of a sunflower is $3\frac{1}{2}$ times the height of the daisy. Find the height of the sunflower.

**21.** Jasmine buys an apple for $0.45, an orange for $0.55, and a pear for $0.99. Write an expression you could use to mentally calculate her total. What is her total?

**22.** The distance from the library to the park is 1.2 miles, and the distance from the park to the pool is 0.5 mile. The park is between the library and the pool. Show that the distance from the library to the pool is the same as the distance from the pool to the library.

**23.** Greeting cards cost $2 each and wrapping paper costs $3 per roll. Write an expression you could use to find the total cost of buying 6 greeting cards and 6 rolls of wrapping paper. What is the total cost?

**Name** _____ **Date** _____

# Integers

Numbers greater than zero are called **positive numbers**. Numbers less than zero are called **negative numbers**. The set of numbers that includes positive and negative numbers, and zero are called **integers**.

---

**EXAMPLE**

*Emily recorded the temperature at noon for a week. The temperatures she recorded were 9°F, 8°F, −6°F, −3°F, −1°F, 2°F, and 1°F. What was the lowest and highest temperature recorded?*

To answer the question, locate the temperatures on a number line.

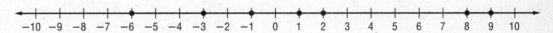

On a number line, values increase as you move to the right.

Since −6 is furthest to the left, −6°F is the coldest temperature. 9 is the farthest number to the right, so 9°F is the highest temperature.

---

The **absolute value** of a number is the positive number of units a number is from zero on a number line.

---

**EXAMPLE**

*Refer to the table. Which city's population changed the most?*

Find the absolute value of each number.

$|+22,457| = 22,457$
$|-84,860| = 84,860$
$|+78,560| = 78,560$
$|-76,704| = 76,704$
$|+49,974| = 49,974$
$|-68,027| = 68,027$

| Population Change, 1990–2000 | |
|---|---|
| Atlanta, GA | +22,457 |
| Baltimore, MD | −84,860 |
| Columbus, OH | +78,560 |
| Detroit, MI | −76,704 |
| Indianapolis, IN | +49,974 |
| Philadelphia, PA | −68,027 |

Since the absolute value of −84,860 is the greatest, Baltimore, Maryland, had the greatest population change.

---

## EXERCISES  *Fill in each blank with <, >, or = to make a true sentence.*

1. 5 ____ −5
2. −4 ____ 3
3. 0 ____ −2
4. −6 ____ −12
5. −35 ____ −16
6. 19 ____ −22
7. 34 ____ 21
8. 23 ____ 23
9. −45 ____ −52

*Write each set of integers in order from least to greatest.*

10. {45, −23, 55, 0, −12, −37}

11. {56, −22, 34, −34, 12, −12}

12. {−450, −100, 254, 564, −356}

13. {1,276, −3,456, −943, −237, −467}

*Find the absolute value.*

14. $|-3|$
15. $|-5|$
16. $|16|$
17. $|27|$
18. $|156|$
19. $|-359|$
20. $|-821|$
21. $|1,436|$

## APPLICATIONS  *Write an integer to describe each situation.*

22. Julio finished the race 3 seconds ahead of the second place finisher.

23. Matthew ended his round of golf 4 under par.

24. Denver is called the Mile High City because its elevation is 5,280 feet above sea level.

*For Exercises 25–27, refer to the table.*

25. Use a number line to order the temperatures from least to greatest.

26. The record low temperature for Michigan is −51°F. Which states have higher record low temperatures?

27. Indiana's record low temperature is −36°F. Which states in the table have lower record low temperatures?

| Record Low Temperatures | |
|---|---|
| California | −45°F |
| Illinois | −36°F |
| Maine | −48°F |
| Nevada | −50°F |
| New York | −52°F |
| Pennsylvania | −42°F |
| Washington | −48°F |

# Adding and Subtracting Integers

You can use a number line to add integers. Locate the first addend on the number line. Move right if the second addend is positive. Move left if the second addend is negative.

---

**EXAMPLE**  *Find 3 + ( −8).*

Start at 3. Since −8 is negative, move left 8 units.

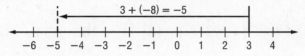

$$3 + (-8) = -5$$

Therefore, 3 + (−8) = − 5.

---

When you add integers, remember:
- The sum of two positive integers is positive.
- The sum of two negative integers is negative.
- The sum of a positive and negative integer is:
  positive if the positive integer has the greater absolute value.
  negative if the negative integer has the greater absolute value.

To subtract an integer, add its opposite.

---

**EXAMPLES**  *Find 4 − 7.*

$$4 - 7 = 4 + (-7) \qquad \text{To subtract 7, add } -7.$$
$$= {}^-3$$

*Find 5 − (−6).*

$$5 - (-6) = 5 + (+6) \qquad \text{To subtract } -6, \text{ add } +6.$$
$$= 11$$

---

## EXERCISES  *Find each sum or difference.*

1. $15 + (-10)$
2. $-20 + (-9)$
3. $16 - (-3)$

4. $-11 - (-6)$
5. $65 - (-45)$
6. $-11 + (-19)$

7. $12 + 15$
8. $-2 - 16$
9. $8 - 17$

10. $16 + (-8)$
11. $-8 + 34$
12. $-12 + (-37)$

13. $23 - 17$
14. $-9 - 25$
15. $14 + 98$

16. $-63 + 53$
17. $(-27) - (-18)$
18. $31 - 74$

19. $81 + 62$
20. $41 - (-35)$
21. $-55 - 23$

22. $20 + (-50)$
23. $-16 - (-16)$
24. $-125 + 79$

## APPLICATIONS  *Great Adventures Outdoor Shop reported profits and losses for a five-month period as shown in the table.*

| Profit and Loss | |
| --- | --- |
| May | profit of $800 |
| June | loss of $1,400 |
| July | loss of $900 |
| August | profit of $500 |
| September | profit of $1,200 |

25. How much more money was lost in June than in July?

26. How much more were the total profits for the last two months than for the first three months?

27. From May through September, did the store have an overall loss or gain and how much?

28. How much did the store lose in October if the overall loss from May through October was $500?

**Name** _____  **Date** _____

# Divisibility Patterns

Sometimes we need to know if a number is divisible by another number. In other words, does a number divide evenly into another number? You can use divisibility patterns to help determine this.

A number is divisible by:
- 2 if the ones digit is divisible by 2.
- 3 if the sum of the digits is divisible by 3.
- 4 if the number formed by the last two digits is divisible by 4.
- 5 if the ones digit is 0 or 5.
- 6 if the number is divisible by 2 and 3.
- 8 if the number formed by the last three digits is divisible by 8.
- 9 if the sum of the digits is divisible by 9.
- 10 if the ones digit is 0.

**EXAMPLE**  *Determine whether 1,080 is divisible by 2, 3, 4, 5, 6, 8, 9, or 10.*

**2:** The ones digit is 0 which is divisible by 2.
So 1,080 is divisible by 2.

**3:** The sum of the digits (1 + 0 + 8 + 0 = 9) is divisible by 3.
So 1,080 is divisible by 3.

**4:** The number formed by the last two digits, 80, is divisible by 4.
So 1,080 is divisible by 4.

**5:** The ones digit is 0.
So 1,080 is divisible by 5.

**6:** The number is divisible by 2 and 3.
So 1,080 is divisible by 6.

**8:** The number formed by the last three digits, 080 or 80, is divisible by 8.
So 1,080 is divisible by 8.

**9:** The sum of the digits (1 + 0 + 8 + 0 = 9) is divisible by 9.
So 1,080 is divisible by 9.

**10:** The ones digit is 0.
So 1,080 is divisible by 10.

1,080 is divisible by 2, 3, 4, 5, 6, 8, 9, and 10.

## EXERCISES
Use the divisibility rules to determine whether the first number is divisible by the second number.

1. 279; 3

2. 1,240; 6

3. 3,250; 5

4. 835; 4

5. 1,249; 8

6. 2,352; 2

Determine whether each number is divisible by 2, 3, 4, 5, 6, 8, 9, or 10.

7. 453

8. 504

9. 672

10. 111

11. 999

12. 3,217

13. 5,200

14. 804

## APPLICATIONS

15. Anna and two of her friends went apple picking. They picked 156 apples. Can they divide the apples equally among themselves? If so, how many apples does each receive?

16. Hannah made 4 dozen muffins. List the ways in which she could package the muffins so that there would be the same number of muffins in each package with no muffins left over.

17. A leap year occurs when the year is divisible by 4. However, years which are divisible by 100 but not divisible by 400 are not leap years. Determine whether the following years are leap years.
    a. 1996
    b. 2004
    c. 1970
    d. 2010
    e. 1776
    f. 1800

# Greatest Common Factor

The **greatest common factor (GCF)** of two or more numbers is the greatest number that is a factor of each number. One way to find the greatest common factor is to list the factors of each number and then choose the greatest common factors.

---

**EXAMPLE**   *Find the GCF of 36 and 48.*

factors of 36:   **1, 2, 3, 4, 6,** 9, **12,** 18, 36
factors of 48:   **1, 2, 3, 4, 6,** 8, **12,** 16, 24, 48

common factors:   1, 2, 3, 4, 6, 12

The GCF of 36 and 48 is 12.

---

Another way to find the GCF is to use the prime factorization of each number. Then identify all common prime factors and find their product.

---

**EXAMPLE**   *Find the GCF of 144 and 180.*

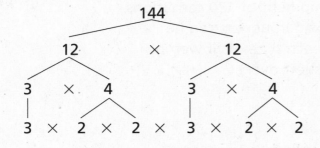
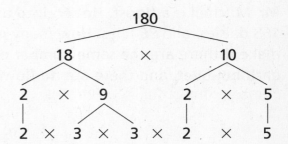

common prime factors:  2, 2, 3, 3

The GCF of 144 and 180 is 2 × 2 × 3 × 3, or 36.

---

*Find the GCF for each set of numbers.*

1. 18, 24

2. 64, 40

3. 60, 75

4. 28, 52

5. 54, 72

6. 48, 72

7. 63, 81

8. 84, 144

9. 72, 170

10. 96, 216

11. 225, 500

12. 121, 231

13. 240, 320

14. 350, 140

15. 162, 243

16. 256, 640

17. 9, 18, 12

18. 30, 45, 15

19. 81, 27, 108

20. 16, 20, 36

21. 98, 168, 196

## APPLICATIONS

22. Sharanda is tiling the wall behind her bathtub. The area to be tiled measures 48 inches by 60 inches. What is the largest square tile that Sharanda can use and not have to cut any tiles?

23. Mr. Mitchell is a florist. He received a shipment of 120 carnations, 168 daisies, and 96 lilies. How many mixed bouquets can he make if there are the same number of each type of flower in each bouquet, and there are no flowers left over?

24. Students at Washington Middle School collected 126 cans of fruit, 336 cans of soup, and 210 cans of vegetables for a food drive. The students are making care packages with at least one of each type of canned good. If the students divide each type of canned good evenly among the care packages, what is the greatest number of care packages if there are no canned goods remaining?

**SKILL 11**

Name _____ Date _____

# Least Common Multiple

A **multiple** of a number is the product of that number and any whole number. The least nonzero multiple of two or more numbers is the **least common multiple (LCM)** of the numbers.

**EXAMPLE**   *Find the least common multiple of 6 and 8.*

positive multiples of 6:    6, 12, 18, **24**, 30, 36, 42, . . .
positive multiples of 8:    8, 16, **24**, 32, 40, 48, 56, . . .

The LCM of 6 and 8 is 24.

**P**rime factorization can also be used to find the LCM.

**EXAMPLE**   *Find the least common multiple of 9, 15, and 21.*

$$9 = 3 \times 3$$
$$15 = 3 \times 5$$
$$21 = 3 \times 7$$

$$3 \times 3 \times 5 \times 7 = 315$$

*Find prime factors of each number.*
*Circle all sets of common factors.*
*Multiply the common factors and any other factors.*

The LCM of 9, 15, and 21 is 315.

**EXERCISES**   *Find the LCM of each set of numbers by listing the multiples of each number.*

1.  3, 4                2.  10, 25                3.  18, 24, 48

*Find the LCM of each set of numbers by writing the prime factorization.*

4.  35, 49                5.  27, 36                6.  10, 12, 15

Glencoe/McGraw-Hill                    21                    Pre-Algebra Intervention

*Find the LCM of each set of numbers.*

**7.** 16, 24       **8.** 56, 16       **9.** 28, 20

**10.** 64, 72       **11.** 63, 77       **12.** 110, 120

**13.** 66, 78, 90       **14.** 40, 60, 108       **15.** 132, 144, 156

**16.** 125, 275, 400       **17.** 196, 225, 256       **18.** 120, 450, 1500

**19.** Find the GCF and LCM of 36 and 54.

**20.** Find the two smallest numbers whose GCF is 7 and whose LCM is 98.

**21.** List the first five multiples of $6p$.

## APPLICATIONS

**22.** Suppose that your taxes, car insurance, and health club membership fees are all due in August. The taxes are due every three months, car insurance is due every six months, and health club membership is due every two months. Name the next month that all three bills will be due in the same month.

**23.** Antoine is buying hamburgers and buns for a class picnic. Hamburgers come in packages of 15 patties and buns come in packages of 8. Antoine wants to have the same number of hamburger patties and buns. What is the least number of hamburger patties and buns he can buy?

**24.** Members of the U.S. House of Representatives are elected every 2 years. United States Senators are elected every 6 years. The President of the United States is elected every 4 years. If a citizen voted for a representative, a senator, and the president in 2004, what is the next year in which the voter can vote for all three in the same year?

**Name** _____ **Date** _____

# Adding and Subtracting Decimals

To add decimals, line up the decimal points. Then add the same way you add whole numbers.

**EXAMPLES**   *16.45 + 18.62*          *77.3 + 88.45 + 90*

$$
\begin{array}{r}
16.45 \\
+\ 18.62 \\
\hline
35.07
\end{array}
$$

$$
\begin{array}{r}
77.30 \\
88.45 \\
+\ 90.00 \\
\hline
255.75
\end{array}
$$

> Annex zeros.

The sum is 35.07.          The sum is 255.75.

To subtract decimals, line up the decimal points. Then subtract the same way you would subtract whole numbers.

**EXAMPLES**   *45.63 − 15.47*          *134 − 105.67*

$$
\begin{array}{r}
45.63 \\
-\ 15.47 \\
\hline
30.16
\end{array}
$$

$$
\begin{array}{r}
134.00 \\
-\ 105.67 \\
\hline
28.33
\end{array}
$$

134.00 ← Annex zeros.

The difference is 30.16.          The difference is 28.33.

**EXERCISES**   *Find each sum or difference.*

1.  $\begin{array}{r} 8.22 \\ +\ 6.83 \\ \hline \end{array}$     2.  $\begin{array}{r} 17.532 \\ -\ 8.173 \\ \hline \end{array}$     3.  $\begin{array}{r} 47.9 \\ +\ 134.2 \\ \hline \end{array}$

4.  $\begin{array}{r} 1.36 \\ -\ 0.48 \\ \hline \end{array}$     5.  $\begin{array}{r} 0.817 \\ -\ 0.6824 \\ \hline \end{array}$     6.  $\begin{array}{r} 68.7 \\ +\ 1.47 \\ \hline \end{array}$

**7.**  46
   − 4.49

**8.**  1.0349
   + 10.08

**9.**  23
   − 4.093

**10.**  47.9 + 32.422

**11.**  52.5 + 8.62

**12.**  3 + 24.15 + 56.052

**13.**  36 + 215.5 + 4.63

**14.**  16.2 − 5.59

**15.**  58 − 0.232

**16.**  23 − 1.59

**17.**  15.6 − 0.423

**18.**  38 + 3.65

**19.**  3.56 + 0.49

**20.**  170 − 67.34

**21.**  43.896 − 22.75

**APPLICATIONS**  *The results of the 2000 presidential election are given at the right. Use this information to answer Exercises 22–24.*

**22.**  What percent of the vote was cast for Bush or Gore?

**23.**  How many more percentage points did Gore receive than Bush?

**24.**  What percent of the vote was cast for listed candidates other than Gore or Bush?

**25.**  Three pieces of cardboard are 0.125 inch, 0.38 inch, and 0.0634 inch thick. What is the combined thickness of all three pieces?

**26.**  A weightlifter lifted 46.8 kilograms on his first lift. His next lift was 50 kilograms. How much more did he lift on his second lift than his first?

**27.**  In a race, the first place finisher had a time of 29.14 seconds. The last-place finisher had a time of 35 seconds. What was the difference between the times?

| 2000 Presidential Elections | |
| --- | --- |
| Candidate | Percent (%) of Popular Vote |
| Browne | 0.36 |
| Buchanan | 0.42 |
| Bush | 47.87 |
| Gore | 48.38 |
| Hagelin | 0.08 |
| Harris | 0.01 |
| Nader | 2.74 |
| Phillips | 0.09 |
| Write-In | 0.02 |
| Other | 0.03 |

**Source:** *The World Almanac*

**Name** _____   **Date** _____

# Multiplying and Dividing Decimals

---

**EXAMPLE**   *Multiply 2.56 by 1.03.*

$$
\begin{array}{r}
2.56 \\
\times\ \ 1.03 \\
\hline
768 \\
000\phantom{0} \\
256\phantom{00} \\
\hline
2.6368
\end{array}
$$

2.56 ←——— 2 decimal places
× 1.03 ←——— 2 decimal places

2.6368 ←——— 4 decimal places

*The sum of the decimal places in the factors is 4, so the product has 4 decimal places.*

The product is 2.6368.

---

**EXAMPLE**   *Divide 0.201 by 0.3.*

$$
\begin{array}{r}
0.67 \\
0.3\overline{)0.2.01} \\
\underline{0\phantom{.....}} \\
2\ 0 \\
\underline{1\ 8} \\
21 \\
\underline{21} \\
0
\end{array}
$$

*Change 0.3 to 3 by moving the decimal point one place to the right.*
*Move the decimal point in the dividend one place to the right.*
*Divide as with whole numbers, placing the decimal point above the new point in the dividend.*

The quotient is 0.67.

---

**EXERCISES**   *Multiply.*

1. 
$$\begin{array}{r} 2.5 \\ \times\ 1.3 \\ \hline \end{array}$$

2. 
$$\begin{array}{r} 6.92 \\ \times\ 53 \\ \hline \end{array}$$

3. 
$$\begin{array}{r} 46.89 \\ \times\ 0.06 \\ \hline \end{array}$$

4. 
$$\begin{array}{r} 925.1 \\ \times\ 30.2 \\ \hline \end{array}$$

5. 
$$\begin{array}{r} 45.21 \\ \times\ 3.2 \\ \hline \end{array}$$

6. 
$$\begin{array}{r} 164.24 \\ \times\ 6.15 \\ \hline \end{array}$$

| 7. | 20.03 | 8. | 10.26 | 9. | 49.76 |
|---|---|---|---|---|---|
| | $\times\ 1.86$ | | $\times\ 30.5$ | | $\times\ 5.17$ |

*Divide.*

10. $0.04\overline{)0.092}$

11. $0.7\overline{)0.245}$

12. $0.06\overline{)0.204}$

13. $0.63\overline{)7.56}$

14. $4.6\overline{)115}$

15. $8.1\overline{)132.03}$

16. $4.7\overline{)43.381}$

17. $0.68\overline{)4.42}$

18. $0.84\overline{)25.62}$

## APPLICATIONS

19. Members of the student body ran 87.75 miles on a 0.25 mile track to raise money for charity. How many laps did they run?

20. A factory manager needs 3.25 yards of material to make a skirt. How many yards of fabric must be used to make 200 skirts?

21. Samantha worked 40.5 hours this week. She makes $9.50 per hour. How much money did she earn this week?

22. Batting averages are calculated to the nearest thousandth. Hikiro has 85 hits in 200 at bats. What is his batting average?

23. Joshua took a 37.5-mile boat trip. It took him 2.5 hours. What was the average speed of the boat?

24. Julia bought 3.5 pounds of mixed nuts that cost $7.49 per pound. How much did 3.5 pounds of nuts cost?

## Adding and Subtracting Fractions

**Name** _____  **Date** _____

To add or subtract fractions with unlike denominators, rename the fractions so that they have a common denominator.

**EXAMPLES** *Find each sum or difference.*

**a.**

$$\frac{1}{4} = \frac{2}{8}$$
$$+\frac{5}{8} = +\frac{5}{8}$$
$$\frac{7}{8}$$

The sum is $\frac{7}{8}$.

**b.**

$$\frac{1}{6} = \frac{5}{30}$$
$$+\frac{7}{10} = +\frac{21}{30}$$
$$\frac{26}{30} = \frac{13}{15}$$

The sum is $\frac{13}{15}$.

**c.**

$$16\frac{1}{2} = 16\frac{7}{14}$$
$$+14\frac{5}{7} = +14\frac{10}{14}$$
$$30\frac{17}{14} = 31\frac{3}{14}$$

The sum is $31\frac{3}{14}$.

**d.**

$$\frac{8}{9} - \frac{1}{3}$$
$$\frac{8}{9} = \frac{8}{9}$$
$$-\frac{1}{3} = -\frac{3}{9}$$
$$\frac{5}{9}$$

The difference is $\frac{5}{9}$.

**e.**

$$\frac{5}{6} - \frac{3}{8}$$
$$\frac{5}{6} = \frac{20}{24}$$
$$-\frac{3}{8} = -\frac{9}{24}$$
$$\frac{11}{24}$$

The difference is $\frac{11}{24}$.

**f.**

$$6 - 3\frac{2}{5}$$
$$6 = 5\frac{5}{5}$$
$$-3\frac{2}{5} = -3\frac{2}{5}$$
$$2\frac{3}{5}$$

The difference is $2\frac{3}{5}$.

**EXERCISES** *Find each sum or difference.*

**1.**
$$\frac{1}{5}$$
$$+\frac{1}{4}$$

**2.**
$$\frac{5}{12}$$
$$+\frac{1}{3}$$

**3.**
$$\frac{1}{6}$$
$$+\frac{3}{5}$$

**4.**
$$\frac{7}{8}$$
$$-\frac{1}{4}$$

**5.**
$$\frac{7}{10}$$
$$-\frac{3}{8}$$

**6.**
$$\frac{11}{12}$$
$$-\frac{1}{6}$$

**7.** $5\frac{1}{4}$
$+\,7\frac{1}{3}$
_____

**8.** $11\frac{3}{4}$
$+\,8\frac{2}{3}$
_____

**9.** $13$
$+\,9\frac{7}{8}$
_____

**10.** $15\frac{1}{2}$
$+\,9\frac{4}{5}$
_____

**11.** $12\frac{1}{2}$
$-\,8\frac{2}{3}$
_____

**12.** $14\frac{5}{8}$
$-\,6\frac{5}{6}$
_____

**13.** $18\frac{7}{8} - 13$

**14.** $11 - 3\frac{5}{9}$

**15.** $16\frac{2}{5} - 13\frac{3}{4}$

**16.** $\frac{3}{10} + \frac{4}{15}$

**17.** $\frac{3}{8} + \frac{5}{12}$

**18.** $18\frac{5}{18} - 8\frac{1}{9}$

**19.** $2\frac{1}{4} + 3\frac{1}{2} + 5\frac{5}{6}$

**20.** $15\frac{3}{4} + 12\frac{5}{16} + 10\frac{3}{8}$

**21.** $21 + 8\frac{7}{10} + 14\frac{3}{4}$

## APPLICATIONS

**22.** Ashley spends $\frac{1}{4}$ of her study time studying math and $\frac{1}{6}$ of her time studying history. How much of her study time does she spend on math and history?

**23.** Hinto repaired her bike for $\frac{5}{6}$ hour and then rode it for $\frac{3}{5}$ hour. How much more time did she spend repairing her bike?

**24.** A tailor buys some cloth to make pants. He buys $3\frac{5}{6}$ yards of one type of fabric and $4\frac{7}{36}$ yards of another. How much fabric did he buy in all?

**25.** A park ranger led a group of campers on a $5\frac{1}{2}$-mile hike. They have already hiked $2\frac{1}{3}$ miles. How far do they have yet to hike?

# Multiplying and Dividing Fractions

To multiply fractions, multiply the numerators and multiply the denominators.

---

**EXAMPLE**   *What is the product of $\frac{5}{6}$ and $\frac{9}{10}$?*

$$\frac{5}{6} \times \frac{9}{10} = \frac{5 \times 9}{6 \times 10} \qquad \text{Multiply the numerators.} \\ \text{Multiply the denominators.}$$

$$= \frac{45}{60} \text{ or } \frac{3}{4} \qquad \text{Simplify.}$$

The product is $\frac{3}{4}$.

---

To divide by a fraction, multiply by its reciprocal.

---

**EXAMPLE**   *What is the quotient of $\frac{4}{15}$ and $\frac{2}{5}$?*

$$\frac{4}{15} \div \frac{2}{5} = \frac{4}{15} \times \frac{5}{2} \qquad \text{Multiply by the reciprocal of } \frac{2}{5} \text{, which is } \frac{5}{2}.$$

$$= \frac{4 \times 5}{15 \times 2} \qquad \text{Multiply the numerators.} \\ \text{Multiply the denominators.}$$

$$= \frac{20}{30} \text{ or } \frac{2}{3} \qquad \text{Simplify.}$$

The quotient is $\frac{2}{3}$.

---

**EXERCISES**   *Multiply. Express each answer in simplest form.*

1. $\frac{2}{3} \times \frac{1}{4}$   2. $\frac{3}{7} \times \frac{1}{2}$   3. $\frac{7}{10} \times \frac{5}{7}$

4. $\frac{5}{8} \times \frac{1}{4}$   5. $\frac{1}{6} \times \frac{3}{5}$   6. $\frac{4}{5} \times \frac{9}{10}$

7. $6 \times \frac{2}{3}$   8. $\frac{3}{5} \times 10$   9. $12 \times \frac{5}{16}$

*Divide. Express each answer in simplest form.*

**10.** $\frac{3}{4} \div \frac{1}{2}$

**11.** $\frac{1}{5} \div \frac{1}{4}$

**12.** $\frac{3}{8} \div \frac{3}{4}$

**13.** $\frac{4}{5} \div \frac{2}{5}$

**14.** $\frac{7}{8} \div \frac{1}{4}$

**15.** $\frac{4}{7} \div \frac{8}{9}$

**16.** $\frac{4}{9} \div \frac{2}{3}$

**17.** $\frac{5}{9} \div 5$

**18.** $20 \div \frac{3}{10}$

*Find each product or quotient. Express each answer in simplest form.*

**19.** $\frac{2}{3} \times \frac{5}{9}$

**20.** $\frac{1}{6} \div \frac{2}{9}$

**21.** $\frac{9}{10} \div \frac{1}{4}$

**22.** $\frac{1}{15} \times 15$

**23.** $\frac{15}{16} \div \frac{15}{16}$

**24.** $\frac{4}{5} \times \frac{15}{24}$

## APPLICATIONS

**25.** A piece of lumber 12 feet long is cut into pieces that are each $\frac{2}{3}$ foot long. How many short pieces are there?

**26.** About $\frac{1}{20}$ of the population of the world lives in South America. If $\frac{1}{35}$ of the population of the world lives in Brazil, what fraction of the population of South America lives in Brazil?

**27.** There is $\frac{1}{3}$ pound of peanuts in 2 pounds of mixed nuts. What part of the mixed nuts are peanuts?

**28.** Three fourths of an apple pie is left over from dessert. If the pie was originally cut in $\frac{1}{16}$ pieces, how many pieces are left?

**29.** A recipe calls for $\frac{1}{8}$ cup of sugar. Christopher is making half the recipe. How much sugar will he need?

**30.** Ms. Valdez has 2 dozen golf balls. She lost $\frac{1}{3}$ of them. How many golf balls does she have left?

**Name** _____ **Date** _____

# Changing Fractions to Decimals

A fraction is another way of writing a division problem. To express a fraction as a decimal, divide the numerator by the denominator. If the division ends, or terminates, with a zero, the decimal is a **terminating decimal.**

---

**EXAMPLE**  *Express $\frac{3}{4}$ as a decimal.*

$\frac{3}{4}$ means $3 \div 4$ or $4\overline{)3}$.

$$
\begin{array}{r}
0.75 \\
4\overline{)3.00} \\
\underline{28\phantom{0}} \\
20 \\
\underline{20} \\
0
\end{array}
$$

*Annex zeros as needed.*

*Division ends when the remainder is 0.*

So, $\frac{3}{4} = 0.75$.

---

If the decimal repeats a pattern in the digits rather than terminates, the decimal is a **repeating decimal.** You can use bar notation to show that a number repeats indefinitely. A bar is written over the digits that repeat.

---

**EXAMPLE**  *Express $\frac{5}{6}$ as a decimal.*

$$
\begin{array}{r}
0.8333 \\
6\overline{)5.0000} \\
\underline{4\,8\phantom{00}} \\
20 \\
\underline{18} \\
20 \\
\underline{18} \\
20 \\
\underline{18} \\
2
\end{array}
$$

*The number 3 repeats.*

*The remainder after each step is 2.*

So, using bar notation, $\frac{5}{6} = 0.8\overline{3}$.

---

*Express each fraction as a decimal. Use bar notation if necessary.*

1. $\frac{3}{5}$     2. $\frac{2}{3}$     3. $\frac{1}{8}$     4. $\frac{2}{9}$

5. $\frac{4}{11}$     6. $\frac{1}{2}$     7. $\frac{3}{10}$     8. $\frac{3}{8}$

9. $\frac{5}{12}$     10. $\frac{4}{9}$     11. $\frac{7}{16}$     12. $\frac{17}{20}$

13. $\frac{1}{6}$     14. $\frac{28}{42}$     15. $\frac{17}{32}$     16. $\frac{13}{25}$

17. $\frac{63}{100}$     18. $\frac{19}{22}$     19. $\frac{37}{50}$     20. $\frac{49}{99}$

21. $\frac{81}{150}$     22. $\frac{267}{500}$     23. $\frac{370}{450}$     24. $\frac{784}{999}$

**APPLICATIONS** *Ms. Breckenridge uses the grading scale shown at the right.*

25. If a student gets $\frac{19}{25}$ of the questions on a quiz correct, what was the student's score?

26. What grade should be given to a student who got 25 out of 30 questions correct if each question was worth the same value?

27. On the first quiz of the grading period, a student answered $\frac{8}{9}$ of the questions correctly. On the second quiz, the student got 22 out of 25 questions correct. Which quiz had the higher score?

| Grade | Score |
|-------|-------|
| A | 93–100 |
| B | 82–92.$\overline{9}$ |
| C | 71–81.$\overline{9}$ |
| D | 60–70.9 |
| F | 0–59.$\overline{9}$ |

# Percents as Fractions and Decimals

To write a percent as a fraction, write a fraction with the percent in the numerator and with a denominator of 100, $\frac{r}{100}$. Then write the fraction in simplest form.

**EXAMPLES** *Express each percent as a fraction.*

a. **40%**

$$40\% = \frac{40}{100}$$

$$= \frac{2}{5}$$

Therefore, $40\% = \frac{2}{5}$.

b. **$87\frac{1}{2}\%$**

$$87\frac{1}{2}\% = \frac{87\frac{1}{2}}{100}$$

$$= \frac{\frac{175}{2}}{100}$$

$$= \frac{175}{2} \times \frac{1}{100}$$

$$= \frac{175}{200}$$

$$= \frac{7}{8}$$

Therefore, $87\frac{1}{2}\% = \frac{7}{8}$.

To express a percent as a decimal, first express the percent as a fraction with a denominator of 100. Then express the fraction as a decimal.

**EXAMPLES** *Express each percent as a decimal.*

a. **51%**

$$51\% = \frac{51}{100}$$

$$= 0.51$$

Therefore, $51\% = 0.51$.

b. **90.2%**

$$90.2\% = \frac{90.2}{100}$$

$$= \frac{90.2 \times 10}{100 \times 10}$$

$$= \frac{902}{1,000}$$

$$= 0.902$$

Therefore, $90.2\% = 0.902$.

*Express each percent as a fraction.*

1. 75%

2. 84%

3. 90%

4. $18\frac{1}{2}\%$

5. 38%

6. $33\frac{1}{3}\%$

7. 56%

8. 60%

*Express each percent as a decimal.*

9. 82%

10. 61.5%

11. 8.9%

12. $48\frac{1}{2}\%$

13. 70%

14. $27\frac{1}{4}\%$

15. 3%

16. 0.25%

*Write each percent as a fraction in simplest form and write as a decimal.*

17. 18%

18. 22%

19. $82\frac{1}{2}\%$

20. $\frac{5}{8}\%$

21. $91\frac{2}{3}\%$

22. 19.6%

23. 0.5625%

24. 4.9%

## APPLICATIONS

25. The average household in the United States spends 15% of its money on food. Express 15% as a decimal.

26. Bananas grow on plants that can be 30 feet tall. A single banana may be 75% water. Express 75% as a fraction and as a decimal.

27. In the United States, showers usually account for 32% of home water use. Express this percent as a fraction and as a decimal.

28. Only 2% of earthquakes in the world occur in the United States. Express this percent as a fraction and as a decimal.

# Percent of a Number

To find the percent of a number, you can either change the percent to a fraction and then multiply, or change the percent to a decimal and then multiply.

---

**EXAMPLE**  *Yankee Stadium in New York has a capacity of about 57,500. If attendance for one baseball game was about 90%, approximately how many people attended the game?*

Change the percent to a decimal.

$90\% = \frac{90}{100}$ or 0.9

Multiply the number by the decimal.

$57,500 \times 0.9 = 51,750$

About 51,750 people attended the game.

---

**EXERCISES**  *Find the percent of each number.*

1. 50% of 48

2. 25% of 164

3. 70% of 90

4. 60% of 125

5. 55% of 960

6. 35% of 600

7. 15% of 120

8. 6% of 50

9. 200% of 13

10. 55% of 84

11. 16% of 48

12. 150% of 60

13. 45% of 80

14. 60% of 40

15. 18% of 300

16. 5% of 16

17. 15% of 50

18. 100% of 47

19. 12.5% of 60

20. 0.02% of 80

**21.** 0.5% of 180

**22.** 0.1% of 770

**23.** 1.4% of 40

**24.** 1.05% of 62

**25.** $12\frac{1}{2}$% of 70

**26.** $5\frac{3}{8}$% of 200

**27.** $2\frac{1}{4}$% of 150

**28.** $33\frac{1}{3}$% of 45

**APPLICATIONS** *Sarah has a part-time job. Each week she budgets her money as shown in the table. Use this data to answer Exercises 29–31.*

| Sarah's Budget | |
|---|---|
| Savings | 40% |
| Lunches | 25% |
| Entertainment | 15% |
| Clothes | 20% |

**29.** If Sarah made $90 last week, how much can she plan to spend on entertainment?

**30.** If Sarah made $105 last week, how much should she plan to save?

**31.** If Sarah made $85 last week, how much can she plan to spend on lunches?

**32.** The population of the U.S. was about 290 million people in 2004. The population of the New York Metropolitan area was about 7.3% of the total. About how many people lived in the New York area in 2004?

**33.** Ninety percent of the seats of a flight are filled. There are 240 seats. How many seats are filled?

**34.** Of the people Joaquin surveyed, 60% had eaten lunch in a restaurant in the past week. If Joaquin surveyed 150 people, how many had eaten lunch in a restaurant in the past week?

**35.** A car that normally sells for $25,900 is on sale for 84.5% of the usual price. What is the sale price of the car?

# Percent Proportion

$\mathsf{Y}$ou can use the percent proportion to solve problems involving percents.

$$\frac{a}{b} = \frac{p}{100}$$    $a$ = part    $b$ = base    $p$ = percent

**EXAMPLES**

*23.4 is what percent of 65?*

The part is 23.4 and the base is 65.

$$\frac{a}{b} = \frac{p}{100}$$

$$\frac{23.4}{65} = \frac{p}{100}$$

$$23.4 \cdot 100 = 65 \cdot p$$
$$2,340 = 65p$$
$$36 = p$$

23.4 is 36% of 65.

*55% of what number is 33?*

The part is 33 and the percent is 55% or $\frac{55}{100}$.

$$\frac{a}{b} = \frac{p}{100}$$

$$\frac{33}{b} = \frac{55}{100}$$

$$33 \cdot 100 = 55 \cdot b$$
$$3,300 = 55b$$
$$60 = b$$

55% of 60 is 33.

**EXERCISES**   *Tell whether each number is the part, base, or percent.*

1. What number is 25% of 20?

2. What percent of 10 is 5?

3. 14% of what number is 63?

4. 7 is what percent of 28?

5. 78% of what number is 50?

6. 72 is 24% of what number?

*Write a proportion for each problem. Then solve. Round answers to the nearest tenth.*

7. What percent of 25 is 5?

8. 9.3% of what number is 63?

9. 30% of what number is 27?

10. 126 is 39% of what number?

11. 61.6 is what percent of 550?   12.   108 is 18% of what number?

13. What percent of 84 is 20?   14.   What percent of 400 is 164?

15. 29.7 is 55% of what number?   16.   18% of 350 is what number?

17. 61.5 is what percent of 600?   18.   72.4 is 23% of what number?

19. What number is 31% of 13?   20.   $33\frac{1}{3}$% of what number is 15?

21. Use a proportion to find $12\frac{2}{3}$% of 462. Round to the nearest hundredth.

22. Use a proportion to determine what percent of 512 is 56. Round to the nearest hundredth.

23. Use a proportion to determine 23% of what number is 81.3. Round to the nearest hundredth.

### APPLICATIONS

24. There are 18 girls and 15 boys in Tyler's homeroom. What percent of Tyler's homeroom are boys? Round to the nearest tenth.

25. If 32% of the 384 students in the eighth grade walk to school, about how many eighth graders walk to school?

26. At North Middle School, 53% of the students are girls. There are 927 students at the school. How many of the students are girls?

27. A ticket company sold 12,500 tickets for a concert. If the tickets sold are about 23.5% of the total number of tickets, how many tickets did the ticket company originally have?

28. Jacquie made 18 out of 24 free throws. What percent of free throws did Jacquie make?

**Name** _____ **Date** _____

# Percent of Change

**A** **percent of change** tells the percent an amount has increased or decreased. When an amount increases, the percent of change is a **percent of increase**.

---

**EXAMPLE**

*According to the U.S. Department of Labor, there were approximately 126,708,000 people employed in 1996. In 2002, there were about 136,485,000 people employed. Find the percent of increase in the number of people employed.*

To find the percent of increase, you can follow these steps.

**1** Subtract to find the amount of change.
136,485,000 − 126,708,000 = 9,777,000        *new − original*

**2** Write a ratio that compares the amount of change to the original amount. Express the ratio as a percent.

$$\text{percent of change} = \frac{\text{amount of change}}{\text{original amount}}$$

$$= \frac{9{,}777{,}000}{126{,}708{,}000} \qquad \textit{Substitution}$$

$$\approx 0.0772$$

The number of people employed increased about 7.72%.

---

**W**hen the amount decreases, the percent of change is a **percent of decrease**. Percent of decrease can be found using the same steps.

---

**EXAMPLE**

*A handheld computer that originally sells for $249 is on sale for $219. What is the percent of decrease of the price of the computer?*

$$249 - 219 = 30 \qquad\qquad \textit{original price − new price}$$

$$\text{percent of change} = \frac{\text{amount of change}}{\text{original amount}}$$

$$= \frac{30}{249} \qquad \textit{Substitution}$$

$$\approx 0.12$$

The percent of decrease in the price of the handheld computer is about 12%.

---

*Find the percent of change. Round to the nearest tenth.*

1. old: $14.50
   new: $13.05

2. old: 237 students
   new: 312 students

3. old: 27.4 inches of snow
   new: 22.8 inches of snow

4. old: 12,000 cars per hour
   new: 14,300 cars per hour

5. old: 2.3 million bushels
   new: 3.1 million bushels

6. old: $119.50
   new: $79.67

7. old: $7,082
   new: $10,189

8. old: 37.5 hours
   new: 42.0 hours

9. old: 74.8 million acres
   new: 67.5 million acres

10. old: 5.7 liters
    new: 4.8 liters

**APPLICATIONS**

11. At the beginning of the day, the stock market was at 10,120.8 points. At the end of the day, it was at 10,058.3 points. What was the percent of change in the stock market value?

12. An auto manufacturer suggests a selling price of $32,450 for its sport coupe. The next year it suggests a selling price of $33,700. What is the percent of change in the price of the car?

13. The U.S. Consumer Price Index in 1990 was 391.4. By 2000 the Consumer Price Index was 515.8. Find the percent of change.

14. During the past school year, there were 2,856 students at Main High School. The next year there were 3,042 students. What was the percent of change?

15. During a clearance sale, the price of a television is reduced from $1,099 to $899 the first week. The next week, the price of the television is lowered to $739. What is the percent of change each week? What is the percent of change from the original price to the final price?

**Name** _____  **Date** _____

# Powers and Exponents

$A$n expression like $3 \times 3 \times 3 \times 3 \times 3$ can be written as a power. A power has two parts, a **base** and an **exponent**. The expression $3 \times 3 \times 3 \times 3 \times 3$ can be written as $3^5$.

---

**EXAMPLE**  *Write the expression $m \cdot m \cdot m \cdot m \cdot m \cdot m$ using exponents.*

The base is *m*. It is a factor 6 times, so the exponent is 6.

$$m \cdot m \cdot m \cdot m \cdot m \cdot m = m^6$$

---

$Y$ou can also use powers to name numbers that are less than one by using exponents that are negative integers. The definition of a negative exponent states that $a^{-n} = \dfrac{1}{a^n}$ for $a \neq 0$ and any integer *n*.

---

**EXAMPLE**  *Write the expression $4^{-3}$ using a positive exponent.*

$$4^{-3} = \frac{1}{4^3}$$

---

**EXERCISES**  *Write each expression using exponents.*

1. $2 \cdot 2 \cdot 2 \cdot 2$

2. $(-3)(-3)(-3)(-3)(-3)$

3. $9$

4. $x \cdot x \cdot x$

5. $c \cdot c \cdot d \cdot d \cdot d \cdot d \cdot d$

6. $8 \cdot a \cdot a \cdot a \cdot b$

7. $(k - 2)(k - 2)$

8. $4 \cdot 4 \cdot 4 \cdot 4 \cdot h \cdot h$

9. $(-w)(-w)(-w)(-w)(-w)$

10. $6 \cdot 6 \cdot 6 \cdot y \cdot y \cdot y \cdot y$

**Evaluate each expression if *m* = 3, *n* = 2, and *p* = −4.**

11. $m^4$

12. $n^6$

13. $3p^2$

**14.** $mn^2$             **15.** $m^2 + p^3$          **16.** $(p + 3)^5$

**17.** $n^2 - 3n + 4$     **18.** $-2mp^2$        **19.** $5(n - 4)^3$

*Write each expression using a positive exponent.*

**20.** $6^{-1}$            **21.** $4^{-3}$           **22.** $(-2)^{-4}$

**23.** $d^{-7}$           **24.** $m^{-5}$         **25.** $3b^{-6}$

**26.** $10^{-2}$          **27.** $\dfrac{1}{x^5}$          **28.** $\dfrac{7}{p^{-4}}$

*Write each fraction as an expression using a negative exponent other than $-1$.*

**29.** $\dfrac{1}{4^5}$         **30.** $\dfrac{1}{3^8}$         **31.** $\dfrac{1}{7^3}$

**32.** $\dfrac{1}{64}$         **33.** $\dfrac{1}{27}$         **34.** $\dfrac{1}{1,000}$

*Evaluate each expression if $a = -2$ and $b = 3$.*

**35.** $5^a$            **36.** $b^{-4}$          **37.** $a^{-3}$

**38.** $(-3)^{-b}$       **39.** $ab^{-2}$         **40.** $(ab)^{-2}$

## APPLICATIONS

**41.** The area of a square is found by multiplying the length of a side by itself. If a square swimming pool has a side of length 45 feet, write an expression for the area of the swimming pool using exponents.

**42.** A molecule of a particular chemical compound weighs one millionth of a gram. Express this weight using a negative exponent.

**43.** A needle has a width measuring $2^{-5}$ inch. Express this measurement in standard form.

# Scientific Notation

A number is expressed in scientific notation when it is written as the product of a factor and a power of ten. The factor must be greater than or equal to 1 and less than 10.

**EXAMPLES** *Express each number in standard form.*

$$8.26 \times 10^5 = 8.26 \times 100,000$$
$$= 826,000$$

$10^5 = 100,000$
*Move the decimal point 5 places to the right.*

$$3.71 \times 10^{-4} = 3.71 \times 0.0001$$
$$= 0.000371$$

$10^{-4} = 0.0001$
*Move the decimal point 4 places to the left.*

*Express each number in scientific notation.*

$$68,000,000 = 6.8 \times 10,000,000$$
$$= 6.8 \times 10^7$$

*The decimal point moves 7 places. The exponent is positive.*

$$0.000029 = 2.9 \times 0.00001$$
$$= 2.9 \times 10^{-5}$$

*The decimal point moves 5 places. The exponent is negative.*

**EXERCISES** *Express each number in standard form.*

1. $7.24 \times 10^3$

2. $1.09 \times 10^{-5}$

3. $9.87 \times 10^{-7}$

4. $5.8 \times 10^6$

5. $3.006 \times 10^2$

6. $4.999 \times 10^{-4}$

7. $2.875 \times 10^{-5}$

8. $6.3 \times 10^4$

9. $4.003 \times 10^6$

10. $1.28 \times 10^{-2}$

*Express each number in scientific notation.*

**11.** 7,500,000

**12.** 291,000

**13.** 0.00037

**14.** 12,600

**15.** 0.0000002

**16.** 0.004

**17.** 60,000,000

**18.** 40,700,000

**19.** 0.00081

**20.** 12,500

*Choose the greater number in each pair.*

**21.** $3.8 \times 10^3$, $1.7 \times 10^5$

**22.** 0.0015, $2.3 \times 10^{-4}$

**23.** 60,000,000, $6.0 \times 10^6$

**24.** $4.75 \times 10^{-3}$, $8.9 \times 10^{-6}$

**25.** 0.00145, $1.2 \times 10^{-3}$

**26.** $7.01 \times 10^3$, 7,000

## APPLICATIONS

**27.** The distance from Earth to the Sun is $1.55 \times 10^8$ kilometers. Express this distance in standard form.

**28.** In 2001, the population of Asia was approximately 3,641,000,000. Express this number in scientific notation.

**29.** A large swimming pool under construction at the Greenview Heights Recreation Center will hold 240,000 gallons of water. Express this volume in scientific notation.

**30.** A scientist is comparing two chemical compounds in her laboratory. Compound A has a mass of $6.1 \times 10^{-7}$ gram, and compound B has a mass of $3.6 \times 10^{-6}$ gram. Which of the two compounds is heavier?

# Variables and Expressions

$A$lgebra is a language of symbols. In algebra, letters, called **variables**, are used to represent unknown quantities. A combination of one or more variables, numbers, and at least one operation is called an **algebraic expression**.

$x - 9$ means $x$ minus 9.

$7m$ means 7 times $m$.

$ab$ means $a$ times $b$.

$\frac{h}{4}$ means $h$ divided by 4.

To **evaluate** an algebraic expression, replace the variable or variables with known values and then use the order of operations.

**EXAMPLE**  *Evaluate $2c - 7 + d$ if $c = 8$ and $d = 5$.*

$$\begin{aligned} 2c - 7 + d &= 2(8) - 7 + 5 && \text{Replace } c \text{ with 8 and } d \text{ with 5.} \\ &= 16 - 7 + 5 && \text{Multiply.} \\ &= 9 - 5 && \text{Subtract.} \\ &= 14 && \text{Add.} \end{aligned}$$

**EXERCISES**  *Evaluate each expression if $x = 9$, $y = 5$, and $z = 2$.*

1. $x + 6$

2. $y - 3$

3. $z + 11$

4. $23 - x$

5. $6z$

6. $14 + y$

7. $4z + 5$

8. $24 - 2x$

9. $3y - 7$

10. $\frac{x}{3}$

11. $\frac{14}{z}$

12. $\frac{xy}{15}$

**13.** $4x - 2y$

**14.** $6z - x$

**15.** $18 - 2x$

**16.** $6y - (x + z)$

**17.** $3x - z$

**18.** $5(y + 7)$

**19.** $2x + y - z$

**20.** $5z - y$

**21.** $4x - (z + 2y)$

**22.** $\frac{2x + 3z}{12}$

**23.** $\frac{7z - y}{x}$

**24.** $\frac{5y - 7}{x}$

**25.** $(11 - 3z) + x + y$

**26.** $7(x - z)$

**27.** $6y - 9z$

**28.** $\frac{xy}{3} - z$

**29.** $\frac{40}{y} + x$

**30.** $\frac{4(x - y)}{z}$

**31.** $3x - 2(y - z)$

**32.** $(14 - 6z) + x$

**33.** $10z - (x + y)$

## APPLICATIONS

**34.** The weekly production costs at Jessica's T-Shirt Shack are given by the algebraic expression $75 + 7s + 12t$ where $s$ represents the number of short-sleeve shirts produced during the week and $t$ represents the number of long-sleeve shirts produced during the week. Find the production cost for a week in which 30 short-sleeve and 24 long-sleeve shirts were produced.

**35.** The perimeter of a rectangle can be found by using the formula $2\ell + 2w$, where $\ell$ represents the length of the rectangle and $w$ represents the width of the rectangle. Find the perimeter of a rectangular swimming pool whose length is 32 feet and whose width is 20 feet.

Name _____ Date _____

# Writing Expressions and Equations

Translating verbal phrases and sentences into algebraic expressions and equations is an important skill in algebra. Key words and phrases play an essential role in this skill.

The first step in translating a verbal phrase into an algebraic expression or a verbal sentence into an algebraic equation is to choose a variable and a quantity for the variable to represent. This is called **defining a variable**.

The following table lists some words and phrases that suggest addition, subtraction, multiplication, and division. Once a variable is defined, these words and phrases will be helpful in writing the complete expression or equation.

| Addition | Subtraction | Multiplication | Division |
|----------|-------------|----------------|----------|
| plus | minus | times | divided |
| sum | difference | product | quotient |
| more than | less than | multiplied | per |
| increased by | subtract | each | rate |
| in all | decreased by | of | ratio |
| together | less | factors | separate |

**EXAMPLES**  *Translate the phrase "three times the number of students per class" into an algebraic expression.*

**Words**   three times the number of students per class

**Variable**   Let *s* represent the number of students per class.

**Expression**  $3s$

*Translate the sentence "The weight of the apple increased by five is equal to twelve ounces." into an algebraic equation.*

**Words**   The weight of the apple increased by five is equal to twelve ounces.

**Variable**   Let *w* represent the weight of the apple.

**Equation**  $w + 5 = 12$

**EXERCISES**  *Translate each phrase into an algebraic expression.*

1.  seven points less than yesterday's score

2.  the number of jelly beans divided into nine piles

3.  the morning temperature increased by sixteen degrees

4.  six times the cost of the old book

5.  two times the difference of a number and eight

*Translate each sentence into an algebraic equation.*

6.  The sum of four and a number is twenty.

7.  Fourteen is the product of two and a number.

8.  Nine less than a number is three.

9.  The quotient of a number and five is eleven.

10. Fifteen less than the product of a number and three is six.

**APPLICATIONS**

11. Sierra purchased an ice cream cone for herself and three friends. The cost was $8. Define a variable and then write an equation that can be used to find how much Sierra paid for each ice cream cone.

12. Nicholas weighed 83 pounds at his most recent checkup. He had gained 9 pounds since his last checkup. Define a variable and then write an equation to find Nicholas' weight at the previous checkup.

13. There are three times as many people at the amusement park today than there were yesterday. Today's attendance is 12,000. Define a variable and then write an equation to find yesterday's attendance.

**Name** _____  **Date** _____

# Simplifying Expressions and Equations

**W**hen an algebraic expression is separated into parts by addition and subtraction signs, each part is called a **term**. The numerical part of a term that contains a variable is called the **coefficient** of the variable. **Like terms** are terms that contain the same variables, such as $3a$ and $7a$ or $9mn$ and $2mn$. A term without a variable is called a **constant**. Constant terms are also like terms. An algebraic expression is in **simplest form** if it has no like terms and no parentheses.

| **EXAMPLE** | **Simplify the expression $x + 5(y + 2x)$.** |
|---|---|

$$
\begin{aligned}
x + 5(y + 2x) &= x + 5(y) + 5(2x) && \textit{Distributive Property} \\
&= x + 5y + 10x && \textit{Multiply.} \\
&= 1x + 5y + 10x && \textit{Identity Property} \\
&= 1x + 10x + 5y && \textit{Commutative Property} \\
&= (1 + 10)x + 5y && \textit{Distributive Property} \\
&= 11x + 5y && \textit{Simplify.}
\end{aligned}
$$

**W**hen solving equations, sometimes it is necessary to simplify the equation by combining like terms before the equation can be solved.

| **EXAMPLES** | **Solve each equation.** |
|---|---|

$$
\begin{aligned}
6a - 2a + 5 &= 17 \\
4a + 5 &= 17 && \textit{Combine like terms.} \\
4a + 5 - 5 &= 17 - 5 && \textit{Subtract 5 from each side.} \\
4a &= 12 && \textit{Simplify.} \\
\frac{4a}{4} &= \frac{12}{4} && \textit{Divide each side by 4.} \\
a &= 3 && \textit{Simplify.}
\end{aligned}
$$

$$
\begin{aligned}
4(2x - 1) &= -6(x + 3) \\
8x - 4 &= -6x - 18 && \textit{Distributive Property} \\
8x - 4 + 6x &= -6x - 18 + 6x && \textit{Add 6x to each side.} \\
14x - 4 &= -18 && \textit{Simplify.} \\
14x - 4 + 4 &= -18 + 4 && \textit{Add 4 to each side.} \\
14x &= -14 && \textit{Simplify.} \\
\frac{14x}{14} &= -\frac{14}{14} && \textit{Divide each side by 14.} \\
x &= -1 && \textit{Simplify.}
\end{aligned}
$$

## EXERCISES  *Simplify each expression.*

**1.** $6y + 9y$

**2.** $-4m + 2m$

**3.** $13v - 9v$

**4.** $7z + 5 - 3z + 2$

**5.** $2p - 11p$

**6.** $3g - 6 + 6$

*Solve each equation.*

**7.** $18p - 2p + 6 = 9 + 5$

**8.** $10b - 4 - 6b = 24 - 4$

**9.** $8n + 6 = 19 + 7n$

**10.** $-3m + 8m = 11 - 4 - 2m$

**11.** $6(3w + 5) = 2(10w + 10)$

**12.** $5(3x + 1) = 2(13x - 3)$

**13.** $3a + 4 - 2a - 7 = 4a + 3$

**14.** $4(8 - 3w) = 32 - 8(w + 2)$

## APPLICATIONS

**15.** Suppose you buy 5 videos that each cost $c$ dollars, a DVD for $30, and a CD for $20. Write an expression in simplest form that represents the total amount spent.

**16.** Malik earned $d$ dollars raking leaves. His friend, Isaiah, earned three times as much. A third friend, Daniel, earned five dollars less than Malik. Write an expression in simplest form that represents the total amount earned by the three friends.

**17.** A rectangle has length $2x - 3$ and width $x + 1$. Write an expression in simplest form that represents the perimeter of the rectangle.

# Solve Equations Involving Addition and Subtraction

To solve an equation means to find a value for the variable that makes the equation true. To solve an equation, you need to get the variable by itself.

*Addition Property of Equality:* If you add the same number to each side of an equation, the two sides remain equal.

**EXAMPLE**  *Solve s − 46 = 12.*

$$s - 46 = 12$$
$$s - 46 + 46 = 12 + 46 \quad \text{Add 46 to each side.}$$
$$s = 58$$

Check:     $s - 46 \overset{?}{=} 12$
           $58 - 46 \overset{?}{=} 2$       *Replace s with 58.*
           $12 = 12 ✓$

The solution is 58.

*Subtraction Property of Equality:* If you subtract the same number from each side of an equation, the two sides remain equal.

**EXAMPLE**  *Solve d + 22 = 60.*

$$d + 22 = 60$$
$$d + 22 - 22 = 60 - 22 \quad \text{Subtract 22 from each side.}$$
$$d = 38$$

Check:     $d + 22 \overset{?}{=} 60$
           $38 + 22 \overset{?}{=} 60$       *Replace d with 38.*
           $60 = 60 ✓$

The solution is 38.

*Solve each equation. Check your solution.*

1. $a - 91 = 20$    2. $1.5 + b = 3$    3. $c - 3.5 = 1.25$

4. $d + 140 = 300$    5. $5.6 + e = 7$    6. $f - 65 = 21$

7. $g + 35 = 62$    8. $h - 12 = 52$    9. $j + 16 = 47$

10. $k - 12 = 13$    11. $16 = m + 9$    12. $n + 16 = 34$

13. $20 + p = 40$    14. $22 = q - 12$    15. $r - 75 = 156$

16. $15.6 + s = 52.1$    17. $312 = t - 64$    18. $u - 71 = 23$

**APPLICATIONS**

19. Alexis sold 170 tickets for her school play. She has 290 tickets remaining. How many tickets were available?

20. Hector owns 87 CDs and DVDs. If he has 41 CDs, how many DVDs does Hector own?

21. Brandon is saving to buy a new computer game that costs $49.98. He still needs to save $21.50. How much has Brandon saved so far?

22. There are 34 students in Ms. Kim's class. Twelve of the students wear braces. How many students do not wear braces?

23. Taylor is downloading files from the Internet. She has transferred 8 of the 18 files she has selected. How many files have yet to be transferred?

24. A recipe calls for $2\frac{1}{2}$ cups of flour. Terrence has $1\frac{1}{3}$ cups available. How much more flour does Terrence need?

Name _____ Date _____

# Solve Equations Involving Multiplication and Division

**D**ivision Property of Equality: If you divide each side of an equation by the same nonzero number, the two sides remain equal.

**EXAMPLE** *Solve 14x = 84.*

$$14x = 84$$
$$\frac{14x}{14} = \frac{84}{84} \qquad \textit{Divide each side by 14.}$$
$$x = 6$$

Check: $$14x = 84$$
$$14 \times 6 \overset{?}{=} 84 \qquad \textit{Replace x with 6.}$$
$$84 = 84 \checkmark$$

The solution is 6.

**M**ultiplication Property of Equality: If you multiply each side of an equation by the same number, the two sides remain equal.

**EXAMPLE** *Solve $15 = \frac{y}{7}$.*

$$15 = \frac{y}{7}$$
$$7(15) = 7\left(\frac{y}{7}\right) \qquad \textit{Multiply each side by 7.}$$
$$105 = y$$

Check: $$15 = \frac{y}{7}$$
$$15 \overset{?}{=} \frac{105}{7} \qquad \textit{Replace y with 105.}$$
$$15 = 15 \checkmark$$

The solution is 105.

## EXERCISES

*Solve each equation. Check your solution.*

**1.** $99 = 3a$

**2.** $0.5b = 3$

**3.** $\dfrac{c}{6} = 12$

**4.** $4 = \dfrac{d}{22}$

**5.** $\dfrac{e}{0.3} = 150$

**6.** $5 = 4f$

**7.** $\dfrac{g}{12} = 16$

**8.** $1.2h = 3.6$

**9.** $19 = \dfrac{j}{0.4}$

**10.** $\dfrac{k}{14} = 39$

**11.** $\dfrac{m}{5} = 16.4$

**12.** $8n = 9.6$

**13.** $1.2p = 2.76$

**14.** $72 = \dfrac{q}{1.8}$

**15.** $9r = 729$

**16.** $21s = 147$

**17.** $18t = 3.6$

**18.** $\dfrac{u}{17} = 3.4$

## APPLICATIONS

**19.** City Center Parking Garage charges $0.75 an hour for parking. How long can Andrew park in the garage if he only has $6 for parking?

**20.** Elena is 5 times older than her youngest brother. Elena is 15 years old. How old is her brother?

**21.** Four friends split the cost of lunch equally. If each person pays $7.50, what is the total cost of lunch?

**22.** A bag of 20 oranges costs $6.99. What is the cost of each orange? Round to the nearest cent.

**23.** The area of a rectangle is 168 square centimeters. If the length of the rectangle is 12 centimeters, what is the measure of the width?

**Name** _____ **Date** _____

# Solve Two-Step Equations

To solve two-step equations, you need to add or subtract first. Then you need to multiply or divide.

**EXAMPLES** *Solve each equation.*

$$6x - 3 = 21$$
$$6x - 3 + 3 = 21 + 3 \qquad \text{Add 3 to each side.}$$
$$6x = 24$$
$$\frac{6x}{6} = \frac{24}{6} \qquad \text{Divide each side by 6.}$$
$$x = 4$$

The solution is 4.

$$\frac{y}{10} + 2.5 = 7.5$$
$$\frac{y}{10} + 2.5 - 2.5 = 7.5 - 2.5 \qquad \text{Subtract 2.5 from each side.}$$
$$\frac{y}{10} = 5$$
$$10\left(\frac{y}{10}\right) = 10(5) \qquad \text{Multiply each side by 10.}$$
$$y = 50$$

The solution is 50.

**EXERCISES** *Solve each equation. Check your solution.*

1. $2a + 7 = 15$

2. $\frac{b}{7} + 10 = 40$

3. $8 - 1.2c = 2$

4. $\frac{d}{7} - 13 = 12$

5. $6e - 12 = 72$

6. $7f + 8.4 = 16.8$

**7.** $\frac{g}{2} + 11 = 16$      **8.** $\frac{h}{0.2} + 0.5 = 10$      **9.** $8 + 5j = 53$

**10.** $50 - 3k = 35$      **11.** $\frac{m}{3} - 5 = 2$      **12.** $6n + 4 = 58$

**13.** $\frac{p}{4} - 2 = 0.8$      **14.** $7q - 9.4 = 11.6$      **15.** $4 = \frac{r}{5} - 16$

**16.** $15 + \frac{s}{8} = 27$      **17.** $8t - 4.6 = 68.2$      **18.** $0.93 = 0.15 + 0.4u$

## APPLICATIONS

**19.** Austin's doctor recommended that he take 4 doses of antibiotics the first day and two doses per day until all the medicine was gone. If the prescription was for 24 doses, how many days did Austin take the medicine?

**20.** A carpet store has carpet for $13.99 per square yard and charges $50 for installation. If a customer paid $364.78, approximately how many square yards of carpet were purchased?

**21.** To convert a temperature in degrees Celsius to degrees Fahrenheit you can use the formula $F = \frac{9}{5}C + 32$. If the outside temperature is 63°F, what is the temperature in degrees Celsius? Round to the nearest whole degree.

**22.** A wireless phone company charges $34.99 a month for phone service. They also charge $0.48 per minute for long distance calls. If Vanessa's bill at the end of the billing period is $64.75, how many minutes of long distance calls did she make?

# Solve Inequalities

**I**nequalities are sentences that compare two quantities that are not necessarily equal. The symbols below are used in inequalities.

| Symbol | Words |
|--------|-------|
| $<$ | less than |
| $>$ | greater than |
| $\leq$ | less than or equal to |
| $\geq$ | greater than or equal to |

**EXAMPLES**  *Solve each inequality. Show the solution on a number line.*

$$2n + 1 > 5$$
$$2n + 1 - 1 > 5 - 1 \qquad \textit{Subtract 1 from each side.}$$
$$2n > 4$$
$$\frac{2n}{2} > \frac{4}{2} \qquad \textit{Divide each side by 2.}$$
$$n > 2$$

To graph the solution on a number line, draw an open circle at 2. Then draw an arrow to show all numbers greater than 2.

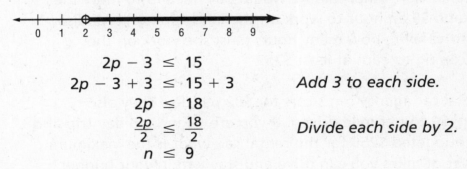

$$2p - 3 \leq 15$$
$$2p - 3 + 3 \leq 15 + 3 \qquad \textit{Add 3 to each side.}$$
$$2p \leq 18$$
$$\frac{2p}{2} \leq \frac{18}{2} \qquad \textit{Divide each side by 2.}$$
$$n \leq 9$$

To graph the solution on a number line, draw a closed circle at 9. Then draw an arrow to show all numbers less than 9.

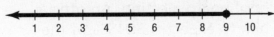

**EXERCISES** *Solve each inequality. Graph the solution on a number line.*

1. $a + 7 < 12$

2. $b - 3 > 8$

3. $2c - 7 \geq 9$

4. $5d + 7 \leq 32$

5. $e + 2 > 16$

6. $f + 12 < 18$

7. $\dfrac{g}{2} \geq 3$

8. $\dfrac{h}{2} + 6 < 8$

9. $\dfrac{j}{3} + 6 \leq 10$

10. $\dfrac{k}{4} + 2 > 3$

**APPLICATIONS**

19. Madison wants to earn at least $75 to spend at the mall this weekend. Her father said he would pay her $15 to mow the lawn and $5 an hour to work on the landscaping. If Madison mows the lawn, how many hours must she work on the landscaping to earn at least $75?

20. A rental car agency rents cars for $32 per day. They also charge $0.15 per mile driven. If you are taking a 5-day trip and have budgeted $250 for the rental car, what is the maximum number of miles you can drive and stay within your budget?

21. Mr. Stamos needs 1,037 valid signatures on a petition to become a candidate for the school board election. An official at the board of elections told him to expect that 15% of the signatures he collects will be invalid. What is the minimum number of signatures he should get to help ensure that he qualifies for the ballot?

# Ratio and Proportion

A **ratio** is a comparison of two numbers by division.

<table>
<tr><td>EXAMPLE</td><td>In a class of 25 students there are 12 girls and 13 are boys. Write the relationship of the number of girls to the number of boys as a ratio.<br><br>The ratio of girls to boys can be written as 12 to 13, 12:13, or $\frac{12}{13}$.</td></tr>
</table>

A **proportion** is a statement that two ratios are equal. In symbols, this can be shown by $\frac{a}{b} = \frac{c}{d}$. The cross products of a proportion, $ad$ and $bc$, are equal.

<table>
<tr><td>EXAMPLE</td><td>Determine if the ratios $\frac{3}{5}$ and $\frac{12}{20}$ form a proportion.</td></tr>
</table>

Find the cross products of $\frac{3}{5} = \frac{12}{20}$.

$$\frac{3}{5} \overset{?}{=} \frac{12}{20} \qquad \text{Write the proportion.}$$
$$3(20) \overset{?}{=} 5(12) \qquad \text{Cross multiply.}$$
$$60 = 60 \qquad \text{Simplify.}$$

So, $\frac{3}{5}$ and $\frac{12}{20}$ form a proportion.

If one term of a proportion is not known, you can use the cross products to set up an equation to solve for the unknown term. This is called **solving the proportion.**

<table>
<tr><td>EXAMPLE</td><td>Solve the proportion $\frac{8}{12} = \frac{x}{15}$.</td></tr>
</table>

$$\frac{8}{12} = \frac{x}{15} \qquad \text{Write the proportion.}$$
$$8(15) = 12(x) \qquad \text{Cross multiply.}$$
$$120 = 12(x)$$
$$\frac{120}{12} = \frac{12(x)}{12} \qquad \text{Divide each side by 12.}$$
$$10 = x$$

**EXERCISES** *Express each ratio as a fraction in simplest form.*

1. 12 pennies to 18 coins

2. 15 bananas out of 25 fruits

3. 32 footballs to 40 basketballs

4. 6 cups to 14 pints

5. 8 clarinets out of 15 instruments

6. 16 tulips out of 24 flowers

7. 12 novels out of 27 books

8. 9 poodles to 12 beagles

*Solve each proportion.*

9. $\frac{a}{12} = \frac{3}{9}$

10. $\frac{8}{b} = \frac{12}{21}$

11. $\frac{24}{36} = \frac{c}{15}$

12. $\frac{27}{6} = \frac{18}{d}$

13. $\frac{7}{8} = \frac{e}{56}$

14. $\frac{27}{36} = \frac{6}{f}$

**APPLICATIONS**

15. If 8 gallons of gasoline cost $11.20, how much would 10 gallons cost?

16. A recipe for punch calls for 4 cups of lemonade for every 6 quarts of fruit juice. How many quarts of fruit juice should Elizabeth use if she has already added 10 cups of lemonade?

17. On a map, the scale is 1 inch equals 160 miles. What is the actual distance if the map distance is $3\frac{1}{2}$ inches?

18. One bag of jelly beans contains 14 red jelly beans. How many red jelly beans would be found in 4 bags of jelly beans?

# Proportional Reasoning

$S$uper Value Grocery has a special on oranges this week. The price is 99¢ for 6 oranges.

**EXAMPLE**  *How many oranges can Daniel buy for $3.30?*

$$\frac{\text{oranges}}{\text{cost (¢)}} \longrightarrow \frac{6}{99} = \frac{x}{330} \longleftarrow \frac{\text{oranges}}{\text{cost (¢)}}$$  *Write a proportion.*

$$(6)(330) = (99)(x)$$  *Cross multiply.*

$$1{,}980 = 99x$$  *Simplify.*

$$\frac{1{,}980}{99} = \frac{99x}{99}$$  *Divide each side by 99.*

$$20 = x$$  *Simplify.*

Daniel can buy 20 oranges.

**EXERCISES**  *Write a proportion to solve each problem. Then solve.*

1.  32 ounces of juice are required to make 2 gallons of punch.
    6 gallons of punch require $n$ ounces of juice.

2.  29 students for every teacher.
    348 students for $t$ teachers.

3.  374 miles driven using 22 gallons of gasoline.
    1,122 miles driven using $g$ gallons of gasoline.

4.  21 bolts connect 3 panels.
    $b$ bolts connect 8 panels.

5.  32 pages for 2 sections of newspaper.
    $p$ pages for 5 sections of newspaper.

6.  $2.49 for 3 bottles of water.
    $8.30 for $w$ bottles of water.

**7.** 3 girls for every 2 boys.
261 girls and $b$ boys.

**8.** 8 packages in 2 cases.
$p$ packages in 7 cases.

**9.** $11.50 earned in one hour.
$d$ earned in 6.5 hours.

**10.** 1.5 inches represents 10 feet.
5 inches represents $x$ feet.

**11.** 18 candy bars in 3 boxes.
900 candy bars in $x$ boxes.

**12.** $\frac{1}{2}$ gallon of paint covers 112 square feet.
$n$ gallons of paint covers 560 square feet.

**APPLICATIONS** *Farmers often express their crop yield in bushels per acre. The table at the right shows Mr. Decker's average yields. Use this data to answer Exercises 13–16.*

| Mr. Decker's Yield (Bushels per acre) | |
|---|---|
| Corn | 98 |
| Soybeans | 48 |
| Wheat | 45 |

**13.** How many bushels of corn should Mr. Decker harvest from 80 acres?

**14.** How many bushels of wheat should Mr. Decker expect from 105 acres?

**15.** If Mr. Decker plants soybeans on 90 acres, how many bushels can he expect to harvest?

**16.** Ms. Holleran harvested 3,815 bushels of corn from 35 acres. Is this yield more or less than Mr. Decker's yield?

**17.** Ms. Galvez paid $150 for 600 square feet of roofing. If she needs 240 square feet more, what is the extra cost?

**18.** A picture measuring 25 centimeters long is enlarged on a copying machine to 30 centimeters long. If the width of the original picture is 15 centimeters, what is the width of the enlarged copy?

# Scale Drawings

A **scale drawing** is used to represent an object that is too large to be drawn or built at actual size.

**EXAMPLE**

*Carlos is drawing plans for a new shopping center. The scale of the drawing is $\frac{1}{2}$ inch equals 5 feet. On the drawing, the front of the shopping center is $18\frac{1}{2}$ inches. What is the actual length of the front of the shopping center?*

Express $\frac{1}{2}$ inch as 0.5 inch and $18\frac{1}{2}$ inches as 18.5 inches. Use the scale 0.5 inch = 5 feet to write a proportion.

$$\frac{\text{drawing}}{\text{actual length}} \longrightarrow \frac{0.5}{5} = \frac{18.5}{x} \longleftarrow \frac{\text{drawing}}{\text{actual length}}$$

$0.5x = (5)(18.5)$    *Cross multiply.*

$0.5x = 92.5$    *Simplify.*

$\dfrac{0.5x}{0.5} = \dfrac{92.5}{0.5}$    *Divide each side by 0.5.*

$x = 185$    *Simplify.*

The actual length of the front of the shopping center is 185 feet.

**EXERCISES**    *On a map, the scale is 1 inch equals 40 miles. For each map distance, find the actual distance.*

1.  $2\frac{1}{2}$ inches

2.  12 inches

3.  $\frac{3}{4}$ inch

4.  $7\frac{1}{4}$ inches

5.  $8\frac{1}{2}$ inches

6.  $4\frac{3}{8}$ inches

**On a blueprint of a new house, the scale is $\frac{1}{4}$ inch equals 2 feet. Find the dimensions of the rooms on the blueprint if the actual measurements of the rooms are given.**

7. 20 feet by $16\frac{3}{4}$ feet

8. 17 feet by $12\frac{3}{4}$ feet

9. $11\frac{1}{2}$ feet by $10\frac{1}{4}$ feet

10. 11 feet by $9\frac{1}{2}$ feet

11. 19 feet by 14 feet

12. $10\frac{3}{4}$ feet by $11\frac{1}{4}$ feet

**APPLICATIONS** *An igloo is a domed structure built of snow blocks by Eskimos. Sometimes several families built a cluster of igloos connected by passageways. Use the scale drawing of such a cluster to answer Exercises 13–17.*

13. What is the actual diameter of each living chamber?

14. What is the actual diameter of the entry chamber?

15. What is the actual diameter of the recreation area?

16. What is the actual diameter of the storage area?

17. Estimate the actual distance from the entry chamber to the back ofthe storage chamber.

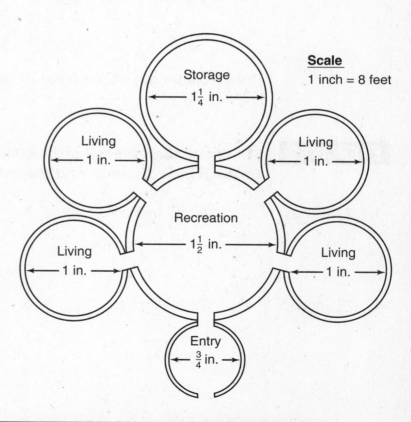

**SKILL 33**

Name _____ Date _____

# Ordered Pairs and the Coordinate Plane

A horizontal number line and a vertical number line meet at their zero points to form a **coordinate plane**. The horizontal line is the **x-axis** and the vertical line is the **y-axis**.

Points are located using ordered pairs. The first number in an ordered pair is the **x-coordinate**, and the second number is the **y-coordinate**.

**EXAMPLES** *Name the ordered pair for point* **A.**

Start at O. Move along the *x*-axis until you are under point A. Then move up until you reach point A. Since you moved 3 units to the left and 4 units up, the ordered pair for point A is (−3, 4).

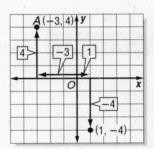

*Graph point (1, −4).*

Start at O. Move 1 unit to the right on the *x*-axis. Then move 4 units down parallel to the *y*-axis to locate the point.

**EXERCISES** *Name the ordered pair for each point.*

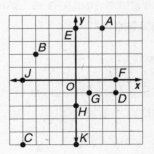

1. A

2. B

3. C

4. D

5. E

6. F

7. G

8. H

9. J

10. K

*Graph and label each point.*

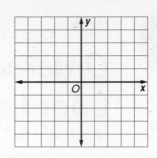

**11.** $M(-4, 2)$      **12.** $N(3, -3)$

**13.** $P(2, 2)$      **14.** $Q(-3, -4)$

**15.** $R(0, -4)$      **16.** $S(-1, 3)$

**17.** $T(-1, -1)$      **18.** $U(3, 4)$

**19.** $W(1, -2)$      **20.** $Z(-4, 0)$

**APPLICATIONS**   *Maps often use a grid system to help locate places on the map. Use the map of Washington D.C. to answer Exercises 21–24.*

**21.** What is located at (B, 1)?

**22.** What is located at (A, 2)?

**23.** Where is the Supreme Court Building located?

**24.** In which section is Union Station Plaza located?

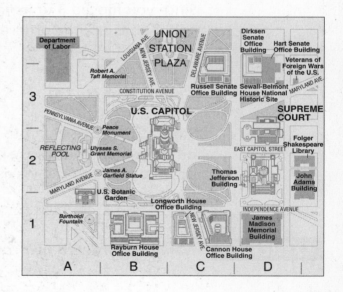

**25.** Use a local map to find an ordered pair to represent each of the following.

     **a.** your house      **b.** your school

     **c.** a friend's house      **d.** your favorite store

**Name** _____     **Date** _____

# Function Tables

$A$ student ticket to the Franklin School of Music's annual concert costs $3.00. The equation that can be used to find the costs of $x$ tickets is $y = 3x$.

**EXAMPLE**   *Make a function table showing the total cost of 2, 4, 6, 8, or 10 tickets.*

$y = 3x$

| $x$ | $3x$ | $y$ |
|-----|------|-----|
| 2 | 3(2) | 6 |
| 4 | 3(4) | 12 |
| 6 | 3(6) | 18 |
| 8 | 3(8) | 24 |
| 10 | 3(10) | 30 |

**EXERCISES**   *Complete each function table.*

**1.** $y = x - 7$

| $x$ | $x - 7$ | $y$ |
|-----|---------|-----|
| 10 | 10 − 7 | 3 |
| 14 | 14 − 7 | |
| 20 | | |
| 25 | | |
| 50 | | |

**2.** $y = x \div 2$

| $x$ | $x \div 2$ | $y$ |
|-----|-----------|-----|
| 4 | | |
| 8 | | |
| 10 | | |
| 30 | | |
| 100 | | |

**3.** $y = 4x - 8$

| $x$ | $4x - 8$ | $y$ |
|-----|----------|-----|
| 5 | 4(5) − 8 | 12 |
| 10 | 4(10) − 8 | |
| 20 | | |
| 50 | | |
| 100 | | |

**4.** $y = 6x + 1$

| $x$ | $6x + 1$ | $y$ |
|-----|----------|-----|
| 2 | | |
| 4 | | |
| 8 | | |
| 20 | | |
| 100 | | |

**5.** $y = 2x - 2$

| x | 2x − 2 | y |
|---|--------|---|
| 1 | | |
| 2 | | |
| 3 | | |
| 4 | | |
| 8 | | |

**6.** $y = 2.5x + 1$

| x | 2.5x + 1 | y |
|---|----------|---|
| 0 | | |
| 2 | | |
| 4 | | |
| 10 | | |
| 25 | | |

**APPLICATIONS** *The cost per hour of operating appliances is listed at the right. Use this information to make a function table for the cost of operating each appliance for 1, 2, 3, 5, or 10 hours.*

| Appliance | Cost per Hour |
|-----------|---------------|
| Television | 12¢ |
| Microwave Oven | 14¢ |
| Vacuum Cleaner | 7¢ |
| Computer | 24¢ |

**7.** Television
$y = 12x$

| x | 12x | y |
|---|-----|---|
| 1 | | |
| 2 | | |
| 3 | | |
| 5 | | |
| 10 | | |

**8.** Microwave Oven
$y = 14x$

| x | 14x | y |
|---|-----|---|
| 1 | | |
| 2 | | |
| 3 | | |
| 5 | | |
| 10 | | |

**9.** Vacuum Cleaner
$y = 7x$

| x | 7x | y |
|---|-----|---|
| 1 | | |
| 2 | | |
| 3 | | |
| 5 | | |
| 10 | | |

**10.** Computer
$y = 24x$

| x | 24x | y |
|---|-----|---|
| 1 | | |
| 2 | | |
| 3 | | |
| 5 | | |
| 10 | | |

**Name** _____    **Date** _____

# Graphing Functions

A function table can be used to graph a function.

**EXAMPLE**

*Chun is riding his bike at an average rate of 14 miles per hour. The function table at the right shows this relationship. Graph the function.*

To graph the function, first label the axes and graph the points named by the data. Then connect the points.

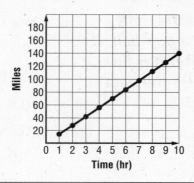

| Time (Hours) | Miles |
|:---:|:---:|
| 1 | 14 |
| 2 | 28 |
| 3 | 42 |
| 4 | 56 |
| 5 | 70 |
| 6 | 84 |
| 7 | 98 |
| 8 | 112 |
| 9 | 126 |
| 10 | 140 |

**EXERCISES**    *Graph each function.*

1.

| Time (min) | Temperature (°C) |
|:---:|:---:|
| 0 | 2 |
| 1 | 5 |
| 2 | 8 |
| 3 | 11 |
| 4 | 14 |
| 5 | 17 |
| 6 | 20 |
| 7 | 23 |
| 8 | 26 |
| 9 | 29 |
| 10 | 32 |

**2.**

| Radius (in.) | Area (sq in.) |
|:---:|:---:|
| 1 | 3.14 |
| 2 | 12.57 |
| 3 | 28.27 |
| 4 | 50.27 |
| 5 | 78.54 |
| 6 | 113.10 |
| 7 | 153.94 |
| 8 | 201.06 |

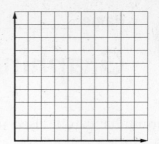

**APPLICATIONS** *The function table at the right shows the height of a golf ball above the ground after it is hit from ground level. Use the data to answer Exercises 3–6.*

| Time (s) | Height (m) |
|:---:|:---:|
| 0 | 0 |
| 0.25 | 4.0 |
| 0.5 | 7.5 |
| 0.75 | 10.25 |
| 1.0 | 12.5 |
| 1.25 | 14.0 |
| 1.5 | 15.0 |
| 1.75 | 15.25 |
| 2.0 | 15.0 |
| 2.25 | 14.0 |
| 2.5 | 12.5 |

**3.** Graph the function.

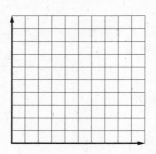

**4.** If the pattern continues, how high above the ground would you expect the golf ball to be after 3.25 seconds?

**5.** Where does the change in the function occur? Why do you think this change occurs?

**6.** How long will it take for the ball to hit the ground?

# Ratios and Rates

A **ratio** is a comparison of two numbers by division. A ratio can be written in several different ways. If there are 5 roses in a bouquet of 12 flowers, then the ratio of roses to total number of flowers in the bouquet can be written as 5 to 12, 5:12, or $\frac{5}{12}$.

---

**EXAMPLE**  **Express the ratio 8 dimes out of 28 coins as a fraction in simplest form.**

$$\frac{8}{28} = \frac{2}{7}$$

The ratio of dimes to coins is 2 to 7. This means that for every 7 coins, 2 of them are dimes.

---

A **rate** is a ratio of two measurements having different kinds of units, such as $25 for 2 dozen. When a rate is simplified so that it has a denominator of 1, it is called a **unit rate**.

---

**EXAMPLE**  **Express the ratio 252 miles in 4 hours as a unit rate.**

$$\frac{252 \text{ miles}}{4 \text{ hours}} = \frac{63 \text{ miles}}{1 \text{ hour}}$$

The unit rate is 63 miles per hour.

---

**EXERCISES**  *Express each ratio as a fraction in simplest form.*

1. 6 strawberries out of 14 pieces of fruit

2. 15 girls to 18 boys

3. 12 blue marbles to 18 green marbles

4. 21 red blocks out of 96 blocks

5. 14 ounces to 35 pounds

6. 15 puppies to 60 kittens

*Express each ratio as a unit rate. Round to the nearest tenth, if necessary.*

7.  $10 for 5 loaves of bread

8.  64 feet in 16 seconds

9.  132 miles on 6 gallons

10. $32 for 5 books

11. 140 meters in 48 seconds

12. 1,400 miles in 4 days

13. $66 for 4 shirts

14. 350 words in 8 minutes

## APPLICATIONS

15. The table below shows the size, in ounces, and the cost of several brands of apple juice. Find the unit cost to determine which brand is the best buy.

| Brand | Size (ounces) | Cost |
| --- | --- | --- |
| Sweeties Apple Juice | 16 | $1.89 |
| Sunshine Apple Juice | 32 | $3.49 |
| Peter's Apple Juice | 64 | $5.09 |

16. A runner training for a marathon ran 18 miles in 150 minutes. Find the length of time it takes the runner to cover 1 mile. Round to the nearest tenth.

17. Alyssa spent $780 on 40 square yards of carpeting for her family room. Find the cost per square yard for the carpet Alyssa selected.

18. During a winter snow storm, a total of 14 inches of snow fell over a period of 8 hours. Find the rate of snowfall per hour. Round to the nearest tenth.

# Slope of a Line

**S**lope describes the steepness of a line. The slope of a line can be expressed as a ratio of the **rise**, vertical change, to the **run**, horizontal change.

$$\text{slope} = \frac{\text{rise}}{\text{run}} \quad \longleftarrow \quad \frac{\text{vertical change}}{\text{horizontal change}}$$

**EXAMPLE**

**Find the slope of the line.**

Choose two points on the line. The points chosen at the right have coordinates (0, 2) and (3, 4).

Draw a vertical line and a horizontal line to connect the points.

Find the length of the vertical segment to find the rise. The rise is 2 units up.

Find the length of the horizontal segment to find the run. The run is 3 units to the right.

$$\text{slope} = \frac{\text{rise}}{\text{run}} = \frac{2}{3}$$

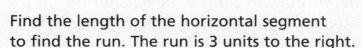

**T**he slope $m$ of a line passing through points at $(x_1, y_1)$ and $(x_2, y_2)$ can be found using the formula $m = \frac{y_2 - y_1}{x_2 - x_1}$, where $x_1 \neq x_2$.

**EXAMPLE**

**Find the slope of the line that passes through A( −5, −3) and B(10, −6).**

Let $A(-5, -3) = (x_1, y_1)$ and let $B(10, -6) = (x_2, y_2)$.

$m = \dfrac{y_2 - y_1}{x_2 - x_1}$      *Definition of slope.*

$= \dfrac{-6 - (-3)}{10 - (-5)}$      $x_1 = -5, y_1 = -3, x_2 = 10, y_2 = -6$

$= \dfrac{-3}{15} \text{ or } \dfrac{1}{-5}$      *Simplify.*

The slope is $\dfrac{1}{-5}$.

## EXERCISES

*Find the slope of each line.*

**1.**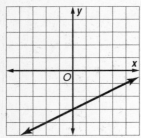

**2.**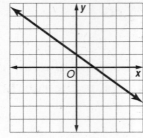

**3.**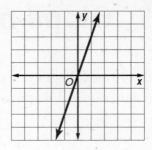

*Find the slope of the line that passes through each pair of points*

**4.** $A(-2, -1)$, $B(3, 9)$

**5.** $C(0, -2)$, $D(3, -3)$

**6.** $E(-5, 20)$, $F(-8, 32)$

**7.** $G(-10, 2)$, $H(10, 8)$

**8.** $J(2, -1)$, $K(6, -11)$

**9.** $M(-3, -14)$ $N(-9, -30)$

## APPLICATIONS

*Paula works as a sales representative for a computer store. She earns a base pay of $1,000 each month. She also earns a commission based on her sales. The graph at the right shows her possible monthly earnings. Use the graph to answer Exercises 10–13.*

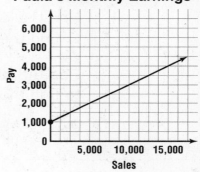

**Paula's Monthly Earnings**

**10.** What is the slope of the line?

**11.** What information is given by the slope of the line?

**12.** If Paula's base pay changed to $1,100, would it change?
**a.** the graph? Why or why not?

**b.** the slope? Why or why not?

**13.** If Paula's rate of commission changed to 25%, would it change the graph? Why or why not?

**Name** _____ **Date** _____

# Graphing Linear Equations

Linear equations can be graphed in the same way that you graph functions.

---

**EXAMPLE**   *Graph the equation y = 3x −2.*

Make a function table for $y = 3x - 2$. Then graph each ordered pair and complete the graph

$$y = 3x - 2.$$

| x | 3x − 2 | y | (x, y) |
|---|--------|---|--------|
| −3 | 3(−3) − 2 | −11 | (−3, −11) |
| −2 | 3(−2) − 2 | −8 | (−2, −8) |
| −1 | 3(−1) − 2 | −5 | (−1, −5) |
| 0 | 3(0) − 2 | −2 | (0, −2) |
| 1 | 3(1) − 2 | 1 | (1, 1) |
| 2 | 3(2) − 2 | 4 | (2, 4) |
| 3 | 3(3) − 2 | 7 | (3, 7) |

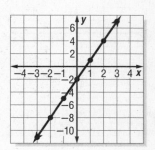

---

**EXERCISES**   *Complete each function table. Then graph the equation.*

**1.** $y = x + 4$

| x | x + 4 | y | (x, y) |
|---|-------|---|--------|
| −2 | | | |
| −1 | | | |
| 0 | | | |
| 1 | | | |
| 2 | | | |

**2.** $y = 6 - 2x$

| x | 6 − 2x | y | (x, y) |
|---|--------|---|--------|
| −2 | | | |
| −1 | | | |
| 0 | | | |
| 1 | | | |
| 2 | | | |

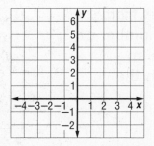

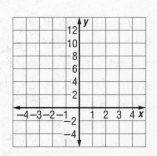

*Graph each equation.*

**3.** $y = -2x$

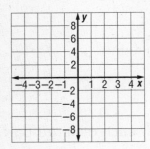

**4.** $y = 3x - 7$

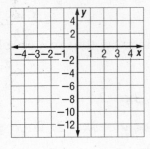

**5.** $y = -x + 6$

**6.** $y = \frac{1}{2}x - 5$

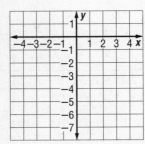

**7.** $y = -\frac{2}{3}x + 2$

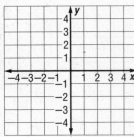

**8.** $y = \frac{4}{3}x + 1$

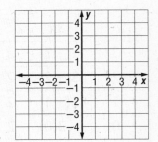

## APPLICATIONS

**9.** A snow storm at Pine Tree Ski Resort deposited $\frac{1}{2}$ foot of snow per hour on top of a 3-foot snow base. Let $x$ represent the number of hours and $y$ represent the total amount of snow. Write an equation to represent the total amount of snow. Graph the equation.

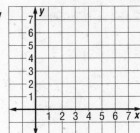

**10.** Alaqua averages 40 miles per hour when she drives from Los Angeles to San Francisco. Let $x$ represent the number of hours and $y$ represent the distance traveled. Write an equation to represent the distance traveled. Graph the equation.

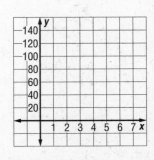

Name _____ Date _____

# Solve Equations in Two Variables

A **linear equation in two variables** is an equation in which the variables appear in separate terms and neither variable contains an exponent other than 1. **Solutions** of a linear equation in two variables are ordered pairs, $(x, y)$ that make the equation true.

**EXAMPLE** *Find four solutions of $y = -3x + 2$. Write the solutions as ordered pairs.*

Choose four values of $x$. Then substitute each value into the equation and solve for $y$.

| $x$ | $y = -3x + 2$ | $y$ | $(x, y)$ |
|-----|---------------|-----|----------|
| $-1$ | $y = -3(-1) + 2$ | $5$ | $(-1, 5)$ |
| $0$ | $y = -3(0) + 2$ | $2$ | $(0, 2)$ |
| $1$ | $y = -3(1) + 2$ | $-1$ | $(1, -1)$ |
| $2$ | $y = -3(2) + 2$ | $-4$ | $(2, -4)$ |

Four solutions are $(-1, 5)$, $(0, 2)$, $(1, -1)$, and $(2, -4)$.

**EXERCISES** *Find four solutions of each equation. Write the solutions as ordered pairs.*

1. $y = x - 3$

2. $y = 2x$

3. $y = 5 - x$

4. $y = 4x - 3$

5. $y = -2x + 4$

**6.** $y = -x$

**7.** $x + y = 5$

**8.** $2x + y = 9$

**9.** $y = -4$

**10.** $x = 3$

## APPLICATIONS

**11.** The equation $y = 3x$ describes the number of eggs ($y$) required to make $x$ batches of brownies. Find the number of eggs required to make 1, 2, 3, and 4 batches of brownies. Express your answers as ordered pairs.

**12.** The equation $y = 3x - 1$ describes the number of employees needed at a restaurant for every 10 customers ($x$). Find the number of employees required for 10, 20, 30, and 40 customers. Express your answers as ordered pairs.

**13.** The equation $y = 4x + 9$ describes the expenses incurred by a pizza shop ($y$) when $x$ pizzas are made. Find the expense for making 4, 5, 6, and 7 pizzas. Express your answers as ordered pairs.

# Square Roots

If $a^2 = b$, then $a$ is the **square root** of $b$. Square roots use the symbol $\sqrt{\phantom{b}}$.

Thus, the square root of $b$ would be written $\sqrt{b}$.

**EXAMPLES**  *Find the square root of each number.*

**36**
Since $6^2 = 36$, $\sqrt{36} = 6$.

**100**
Since $10^2 = 100$, $\sqrt{100} = 10$.

Numbers like 4, 9, 25, and 49 are called **perfect squares** because their square roots are whole numbers.

You can find an estimate for numbers that are not perfect squares.

**EXAMPLE**  *Estimate $\sqrt{95}$ to the nearest whole number.*

The closest perfect square less than 95 is 81.
The closest perfect square greater than 95 is 100.

$$81 < 95 < 100$$
$$\sqrt{81} < \sqrt{95} < \sqrt{100}$$
$$\sqrt{9^2} < \sqrt{95} < \sqrt{10^2}$$
$$9 < \sqrt{95} < 10$$

So, $\sqrt{95}$ is between 9 and 10. Since 95 is closer to 100 than to 81, the best whole number estimate for $\sqrt{95}$ is 10

**EXERCISES**  *Find each square root.*

1. $\sqrt{25}$     2. $\sqrt{49}$     3. $\sqrt{16}$     4. $\sqrt{196}$

5. $\sqrt{256}$     6. $\sqrt{121}$     7. $\sqrt{225}$     8. $\sqrt{484}$

9. $\sqrt{529}$     10. $\sqrt{144}$     11. $\sqrt{576}$     12. $\sqrt{900}$

13. $\sqrt{39}$      14. $\sqrt{106}$      15. $\sqrt{71}$      16. $\sqrt{30}$

17. $\sqrt{15}$      18. $\sqrt{155}$      19. $\sqrt{200}$      20. $\sqrt{250}$

21. $\sqrt{500}$      22. $\sqrt{297}$      23. $\sqrt{340}$      24. $\sqrt{422}$

25. $\sqrt{803}$      26. $\sqrt{644}$      27. $\sqrt{975}$      28. $\sqrt{2,018}$

**APPLICATIONS**    *The area of the floor of a square room is 324 square feet. Use this information to answer Exercises 29–31.*

29. What is the length of each side of the floor?

30. If a square carpet with an area of 144 square feet is placed in the center of the room, what is the width of the floor that is uncovered on each side of the carpet?

31. How many 9-inch square tiles would be required to cover the entire floor?

32. The area of a square is 1,225 square centimeters. What is the perimeter of the square?

33. A bag of Super Green Lawn Fertilizer covers 9,500 square feet. What is the largest square lawn that can be fertilized using one bag of fertilizer? Round to the nearest foot.

34. Trees in orchards are often planted evenly spaced apart in square plots. How many rows of trees are in a plot that contains 1,024 trees?

35. A square playground has an area of 750 square meters. Approximately how much fencing would be required to enclose the playground?

# Sequences

A **sequence** of numbers is a list in a specific order. The numbers in the sequence are called **terms.**

A sequence is an **arithmetic sequence** if you can find the next term by adding the same number to or subtracting the same number from the previous term.

---

**EXAMPLE**   *Find the next three terms in the sequence 7, 11, 15, 19, 23, ....*

7,    11,    15,    19,    23,    ...

+4     +4     +4     +4

Since the difference between each term is the same, the sequence is arithmetic. The pattern is to add 4 to each term. To find the next three terms, add 4 to each term.

23 + 4 = 27      27 + 4 = 31      31 + 4 = 35

The next three terms in the sequence are 27, 31, and 35.

---

A sequence is a **geometric sequence** if you can find the next term by multiplying the previous term by the same number.

---

**EXAMPLE**   *Find the next three terms in the sequence 576, 288, 144, 72, . . ..*

576,   288,   144,   72,    ...

× 0.5   × 0.5   × 0.5

Since each term is found by multiplying the previous term by the same number, the sequence is geometric. The pattern is to multiply each term by 0.5.

72 × 0.5 = 36      36 × 0.5 = 18     18 × 0.5 = 9

The next three terms in the sequence are 36, 18, and 9.

---

*Identify the pattern in each sequence. Then find the next three terms in each sequence.*

**1.** 4, 11, 18, 25, 32, …

**2.** 729, 243, 81, 27, …

**3.** 1, 4, 8, 13, 19, …

*Identify each sequence as arithmetic, geometric, or neither. Then find the next three terms in each sequence.*

**4.** 89, 86, 83, 80, …

**5.** 5, 25, 125, 625, …

**6.** 7, 12, 17, 22, …

**7.** 78, 75, 77, 74, 76, …

**8.** 64, 32, 16, 8, …

**9.** 90, 85, 79, 72, …

**10.** Find the eighth term in the sequence −32, −21, −10, 1, …

**11.** Find the ninth term in the sequence 17, 22, 27, …

**12.** The distances traveled each second by a moving object are given in the table. What kind of sequence is this? What is the pattern?

| Second | Distance Traveled (feet) |
|--------|--------------------------|
| 1 | 16 |
| 2 | 48 |
| 3 | 80 |
| 4 | 112 |

**13.** Joseph plans to open a savings account with $100 from his March paycheck and then increase the amount he deposits by $100 each month. How many months will it take Joseph to have $600 in deposits in his account?

**14.** An apartment rents for $800 a month. The monthly rent is expected to increase $25 each year. How much will the monthly rent be at the end of 4 years?

**15.** Taylan's math average increases one half of a point each week for 10 consecutive weeks. If his average was originally 80, what was his average at the end of 9 weeks?

# Geometric Terms

**A** **point** is a specific location in space with no size or shape. Points are named using capital letters and are represented by a dot.

• A

A **line** is a never-ending straight path extending in two directions. A line can be named by using two points through which it passes, or by a single letter. The line shown could be named $\overleftrightarrow{BC}$ or $\ell$.

A **plane** is a two-dimensional flat surface that extends in all directions. A plane can be named by any three points not on the same line, such as plane *PRQ*, or by a single letter like plane *M*.

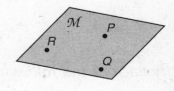

A **line segment** is part of a line containing two endpoints and all points in between. A line segment is named by its endpoints. Line segment *MN*, $\overline{MN}$, is shown.

A **ray** is a part of a line that extends indefinitely from one point in one direction. A ray is named by its endpoint and a second point on the line. Ray *AB*, $\overrightarrow{AB}$ is shown.

---

**EXAMPLES** *Identify each of the following.*

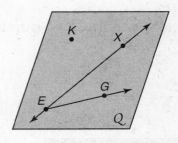

Point
*E, G, K,* and *X* are all points.

Line
$\overleftrightarrow{EX}$

Ray
*EG, EX,* and *XE*

Plane
plane *Q*
The plane can be identified by points it contains such as plane *EGX*, or plane *EGK*.

*Solve each equation. Check your solution.*

1.  Give three names for the line.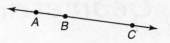

2.  Name two rays and three segments in the figure.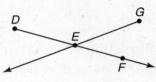

3.  Give three names for the plane.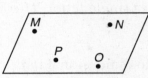

*Identify each of the following.*

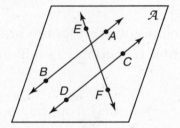

4.  two points

5.  two line segments

6.  a ray and a line

7.  a plane and two points

*Determine whether each model suggests a point, a line, or a plane.*

8.  violin string

9.  meeting of two walls

10. corner of a book

11. ice on an ice rink

12. end of a thumbtack

13. cover of a CD

14. edge of a napkin

15. bulletin board

Name _____ Date _____

# Angles

An angle is formed by two rays with a common endpoint called the **vertex**. Angles are often measured in **degrees** and can be classified according to their measure.

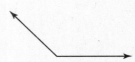

**Right angles**
measure 90°.

**Acute angles**
measure between
0° and 90°.

**Obtuse angles**
measure between
90° and 180°.

A protractor may be used to measure an angle.

---

**EXAMPLE**  *Measure each angle. Classify each angle as right, acute, or obtuse.*

To measure an angle, place the center of the protractor on the vertex of the angle. Place the zero mark of the scale along one side of the angle. Find the place where the other side crosses the scale and read the measure.

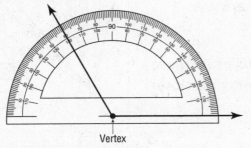

Vertex

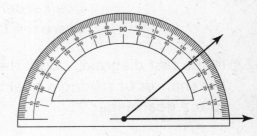

The measure of the angle is 120°. Therefore, it is obtuse.

The measure of the angle is 40°. Therefore, it is acute.

---

## EXERCISES

Use the figure at the right to find the measure of each angle. Then classify each angle as right, acute, or obtuse.

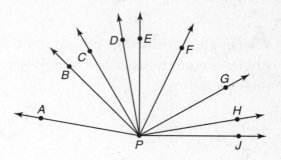

1. ∠FPJ

2. ∠DPJ

3. ∠BPJ

4. ∠EPJ

5. ∠BPF

6. ∠APG

7. ∠DPH

8. ∠EPG

Use a protractor to draw angles having the following measures.

9. 80°

10. 125°

11. 90°

## APPLICATIONS

Soccer is a game of angles. Study the diagrams of a player shooting against a goalkeeper to answer Exercises 12–14.

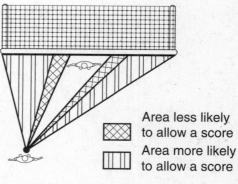

Area less likely to allow a score

Area more likely to allow a score

12. In the first diagram, would the player increase her chances of scoring a goal by moving closer to the goalkeeper?

13. The player forces the goalkeeper to the left as shown in the second diagram. Will the player have a better chance of making a goal?

14. If the goalkeeper moves closer to the player, will the player have a better chance of making a goal?

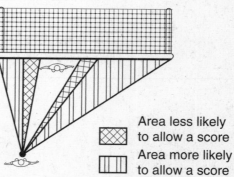

Area less likely to allow a score

Area more likely to allow a score

# Angle Relationships

When two lines intersect, they form two pairs of opposite angles called **vertical angles.** Vertical angles have the same measure and are therefore congruent.

---

**EXAMPLE**   **Find m∠1.**

Angle 1 and the angle whose measure is 132° are vertical angles. Therefore, they are congruent.

Thus, $m\angle 1 = 132°$.

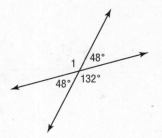

---

Two angles are **complementary** if the sum of their measures is 90°. Two angles are **supplementary** if the sum of their measures is 180°.

---

**EXAMPLES**   **Find x in each figure.**

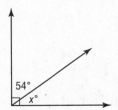

The two angles form a right angle, which measures 90°. Therefore, the angles are complementary.

$$54 + x = 90$$
$$54 + x - 54 = 90 - 54$$
$$x = 36$$

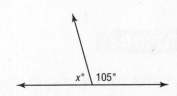

The two angles form a straight line, which measures 180°. Therefore, the angles are supplementary.

$$x + 105 = 180$$
$$x + 105 - 105 = 180 - 105$$
$$x = 75$$

---

## EXERCISES  *Find the value of x in each figure.*

**1.**

**2.**

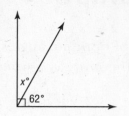

**3.**

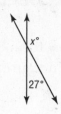

**4.**

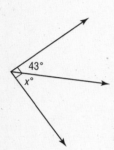

**5.**

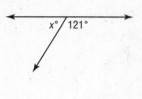

**6.**

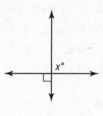

**7.** Angles *A* and *B* are vertical angles. If $m\angle A = 63°$ and $m\angle B = (x + 15)°$, find the value of *x*.

**8.** Angles *P* and *Q* are supplementary angles. If $m\angle P = (x - 25)°$ and $m\angle Q = 102°$, find the value of *x*.

**9.** Angles *Y* and *Z* are complementary. If $M\angle Y = (4x + 2)°$ and $m\angle Z = (5x - 2)°$, find the value of *x*.

## APPLICATIONS

**10.** A carpenter uses a power saw to cut a piece of lumber at a 135° angle. What is the measure of the other angle formed by the cut?

**11.** Megan is making a quilt using the pattern shown at the right.

**a.** What is $m\angle 1$?

**b.** What is $m\angle 2$?

**c.** What is $m\angle 3$?

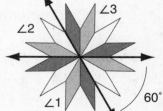

**Name** _____  **Date** _____

# Parallel Lines & Angle Relationships

Two lines in a plane that never intersect or cross are called **parallel lines.** Arrowheads on the lines indicate that they are parallel.

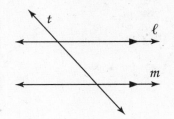

Lines ℓ and *m* are parallel. This can be written as ℓ ∥ *m*. Line *t* is a transversal.

A line that intersects two or more parallel lines is called a **transversal.**

When two parallel lines are cut by a transversal, then the following are true.

- **Alternate interior angles,** those on opposite sides of the transversal and inside the other two lines are congruent. In the figure, ∠2 ≅ ∠7 and ∠4 ≅ ∠5.
- **Alternate exterior angles,** those on opposite sides of the transversal and outside the other two lines, are congruent. In the figure, ∠3 ≅ ∠6 and ∠1 ≅ ∠8.
- **Corresponding angles,** those in the same position on the two lines in relation to the transversal, are congruent. In the figure, ∠1 ≅ ∠5, ∠2 ≅ ∠6, ∠3 ≅ ∠7, and ∠4 ≅ ∠8.
- **Consecutive interior angles,** those on the same side of the transversal and inside the other two lines, are supplementary. In the figure, ∠2 and ∠5 are supplementary, and ∠4 and ∠7 are supplementary.

---

**EXAMPLES**   *Find the measure of each angle.*

If *m*∠4 = 115°, *m*∠6 = _____.
Since ∠4 and ∠6 are alternate interior angles, they are congruent. So, *m*∠6 = 115°

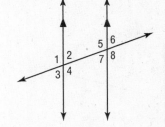

If *m*∠7 = 50°, *m*∠1 = _____.
Since ∠7 and ∠1 are alternate exterior angles, they are congruent. So, *m*∠1 = 50°

If *m*∠1 = 62°, *m*∠5 = _____.
Since ∠1 and ∠5 are corresponding angles, they are congruent.  So, *m*∠5 = 62°

If *m*∠3 = 34°, *m*∠6 = _____.
Since ∠3 and ∠6 are consecutive interior angles, they are supplementary.

$$34 + m∠6 = 180 \qquad \textit{Definition of supplementary angles}$$
$$m∠6 = 146 \qquad \textit{Subtract 34 from each side.}$$

---

1. $m\angle 1 = 42°$, $m\angle 7 =$ _____

2. $m\angle 3 = 58°$, $m\angle 5 =$ _____

3. $m\angle 2 = 110°$, $m\angle 5 =$ _____

4. $m\angle 6 = 127°$, $m\angle 4 =$ _____

5. $m\angle 8 = 150°$, $m\angle 2 =$ _____

6. $m\angle 7 = 60°$, $m\angle 3 =$ _____

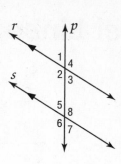

**Find the value of x in each figure.**

7.

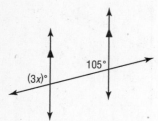

8.

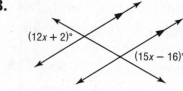

**APPLICATIONS**

9. Brooke is building a bench to place in her yard. The top of the bench will be parallel to the ground. If $m\angle 1 = 135°$, find $m\angle 2$ and $m\angle 3$.

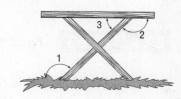

10. A section of fencing is reinforced with a diagonal brace as shown. What is $m\angle ACD$?

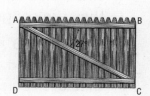

11. City planners angled the parking spaces at City Hall. All of the lines marking the parking spaces are parallel. If $m\angle 2 = 40°$, what is $m\angle 1$? Explain.

# Triangles and Quadrilaterals

**A** triangle is a polygon with three angles and three sides. Triangles may be classified by the measures of their angles or by the lengths of their sides.

| Triangles | | | |
|---|---|---|---|
| **Classification by Angles** | | **Classification by Sides** | |
| Acute | all angles acute | Scalene | all sides different lengths |
| Obtuse | one obtuse angle | Isosceles | at least two sides the same length |
| Right | one right angle | Equilateral | three sides the same length |

**A** quadrilateral is a polygon with four angles and four sides. Sides and angles are also used to classify quadrilaterals.

| Quadrilaterals | |
|---|---|
| Trapezoid | only one pair of parallel sides |
| Parallelogram | both pairs of opposite sides parallel |
| Rectangle | parallelogram with four right angles |
| Rhombus | parallelogram with four sides the same length |
| Square | parallelogram with four right angles and four sides the same length |

**EXAMPLE**    *Identify each polygon.*

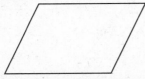

There are two pairs of opposite parallel sides. This quadrilateral is a parallelogram.

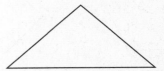

One of the angles is obtuse and two of the sides are the same length. This triangle is obtuse and isosceles.

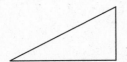

One of the angles is a right angle and none of the sides are the same length. This triangle is right and scalene.

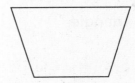

One pair of sides is parallel. This quadrilateral is a trapezoid.

**EXERCISES** *Classify each triangle by its angles and by its sides.*

1.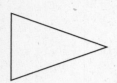

2.

3.

4.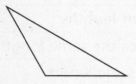

5.

6.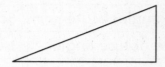

*Name every quadrilateral that describes each figure. Then state which name best describes the figure.*

7.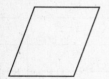

8.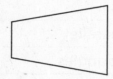

9.

10.

11.

12.

**APPLICATIONS** *Name two real-world examples of each figure.*

13. equilateral triangle

14. trapezoid

15. rectangle

16. right scalene triangle

17. obtuse triangle

18. rhombus

19. square

20. acute isosceles triangle

**Name** _____ **Date** _____

# Similar Figures

**F**igures that have the same shape but not necessarily the same size are similar. You can use ratios to determine whether two figures are similar.

---

**EXAMPLE**   *Determine if the triangles are similar.*

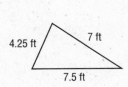
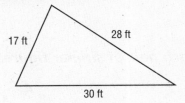

Write ratios comparing the sides of
one triangle to the corresponding
sides of the other triangle.

$\dfrac{\text{side measure of first triangle}}{\text{side measure of second triangle}}$   $\dfrac{4.25}{17} = \dfrac{1}{4}$   $\dfrac{7}{28} = \dfrac{1}{4}$   $\dfrac{7.5}{30} = \dfrac{1}{4}$

The ratios of the corresponding sides all equal $\dfrac{1}{4}$.
Therefore, the triangles are similar.

---

**P**roportions can be used to determine the measures of the sides of similar figures.

---

**EXAMPLE**   *The pentagons are similar. Find the value of x.*

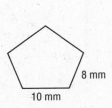
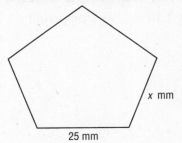

$\dfrac{10}{25} = \dfrac{8}{x}$         *Write a proportion.*

$(10)(x) = (25)(8)$      *Cross multiply.*

$10x = 200$          *Simplify.*

$\dfrac{10x}{10} = \dfrac{200}{10}$        *Divide each side by 10.*

$x = 20$           *Simplify.*

---

## EXERCISES  Determine if each pair of figures is similar.

**1.**

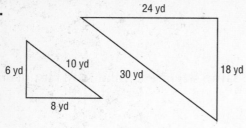

**2.**

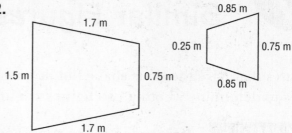

## Find the value of x in each pair of similar figures.

**3.**

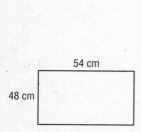

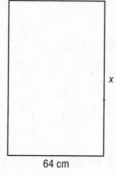

**4.**

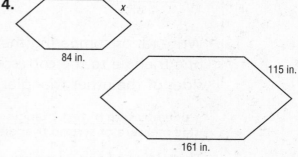

## APPLICATIONS

**5.** A flagpole casts a shadow 5.6 meters long. Isabel is 1.75 meters tall and casts a shadow 0.8 meter long. How tall is the flagpole?

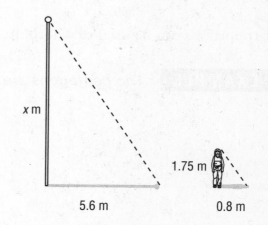

**6.** Will and Kayla want to know how far it is across a pond. They made the sketch at the right. How far is it across the pond?

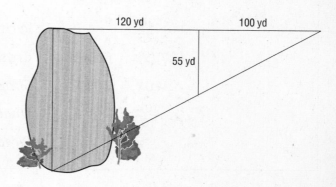

**Name** _____  **Date** _____

# Congruent Figures

Two or more figures that are the same shape and same size are **congruent figures**.

**EXAMPLES**   *Determine if each pair of figures is congruent.*

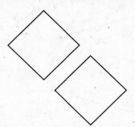

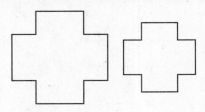

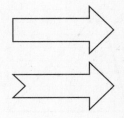

The figures are the same size and the same shape. The figures are congruent.

The figures are the same shape, but not the same size. The figures are not congruent.

The figures are the same size, but not the same shape. The figures are not congruent.

**EXERCISES**   *Determine if each pair of figures is congruent. Explain.*

1.

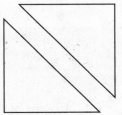

2.

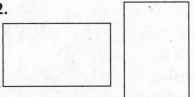

3.

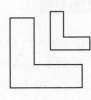

4.

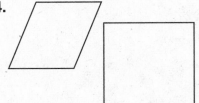

5.

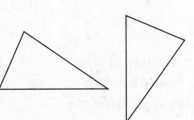

6.

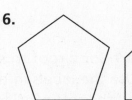

**7.** List the congruent figures.

a.   b.   c.   d.   e.   f.

*In Exercises 8–11, use the grid to draw a figure that is congruent to the given figure.*

**8.**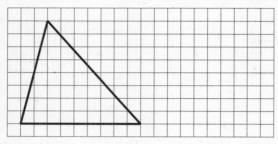

**9.**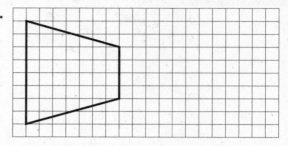

**10.**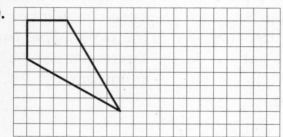

**11.**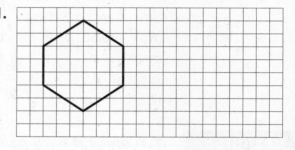

## APPLICATIONS

**12.** Suni is dividing her back yard into two equal-sized gardens with a walkway dividing them. If she wants to put a fence around the outside of Garden 2, how much fencing material will she need.

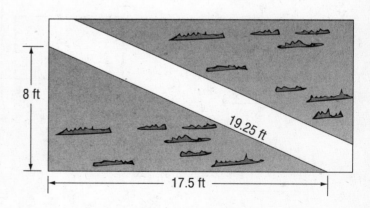

**13.** A pattern for a roof truss is shown at the right. Use the labels in the figures and name all the sets of triangles that appear to be congruent.

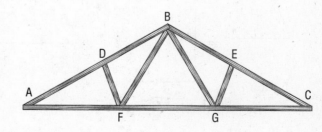

**Name** _____ **Date** _____

# Translations

**A** **translation** is a slide or movement of a figure from one place to another.

**EXAMPLE**    *Translate triangle XYZ 4 units to the left and 2 units up.*

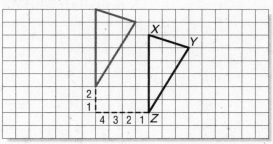

- Move point X 4 units to the left and 2 units up.
- Move point Y 4 units to the left and 2 units up.
- Move point Z 4 units to the left and 2 units up.
- Connect the points to draw the new triangle.

**EXERCISES**    *Translate each figure as indicated.*

**1.** 4 units right and 5 units down

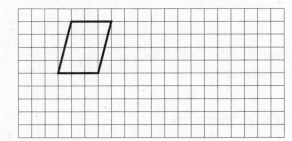

**2.** 6 units left and 3 units up

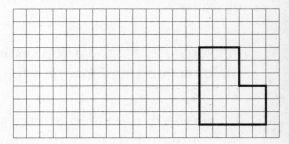

**3.** 7 units left

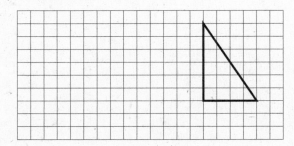

**4.** 10 units left and 2 units up

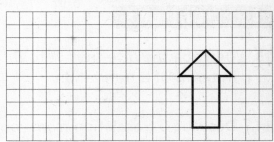

**5.** 3 units right and 5 units down

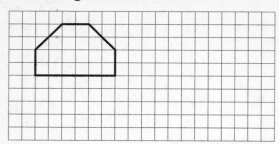

**6.** 8 units left and 4 units up

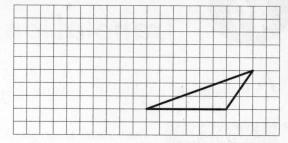

**7.** 2 units right and 3 units down

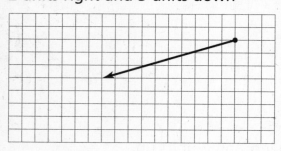

**8.** 5 units left and 6 units up

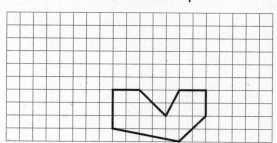

*Answer each question.*

9. Are the translated figures in Exercises 5–8 congruent or similar to the original figures?

10. In Exercise 6, are the angles of the triangle in the same positions?

11. In Exercise 7, are the arrows pointing in the same direction? Is the direction the same for a figure and its translation?

12. In Exercise 8, the indentation is on the top of the figure. In the translation, is the indentation on the top of the figure?

## APPLICATIONS

13. Describe the dive from *A* to *B* in terms of a translation.

14. Describe a translation from your house to a friend's house.

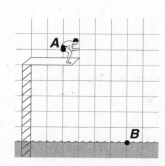

**Name** _____ **Date** _____

# Reflections

$\mathsf{A}$ **reflection** is a transformation that flips a figure over a line of reflection.

**EXAMPLE** *Use the grid to reflect the figure over the given line.*

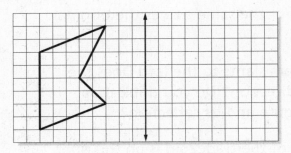

For each vertex on the figure, find the point that is exactly the same distance from the line of reflection, but on the other side of the line. Draw the completed image.

Figure          Reflection

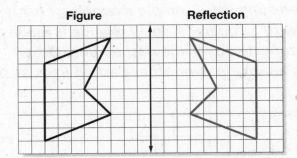

**EXERCISES** *Reflect each figure over the given line.*

1.

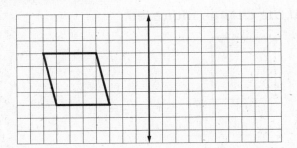

2.

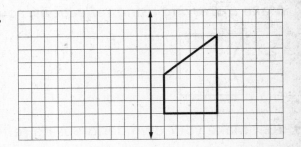

**3.**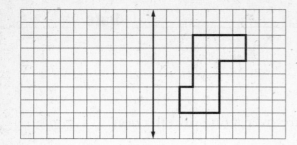

**4.**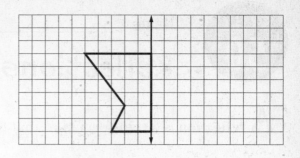

*Refer to the reflections in exercises 1–4.*

**5.** Is a reflection larger, smaller, or the same size as the original figure?

**6.** In Exercise 1, the parallelogram slants to the left. Does the reflection slant in the same direction? Do you think that direction is the same for a figure and its reflection?

**7.** In Exercise 2, the larger base of the trapezoid is on the right. Is the larger base on the right in the reflection?

**APPLICATIONS** *M.C. Escher used transformations such as reflections to create interesting art. A simple example of his type of art starts with a square. A small change is made and this change is reflected over the dashed line. Other reflections are made over other dashed lines as shown.*

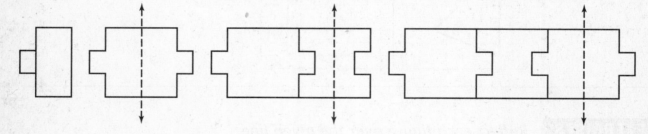

*Make a drawing using reflections, squares, and the change indicated.*

**8.**

**9.**

**10.** Make your own design using reflections.

Name _____ Date _____

# The Pythagorean Theorem

In a right triangle, the sides that form a right angle are called **legs**. The side opposite the right angle is the **hypotenuse**.

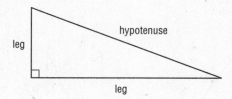

The **Pythagorean Theorem** describes the relationship between the lengths of the legs and the hypotenuse. This theorem is true for any right triangle.

The Pythagorean Theorem states that if a triangle is a right triangle, then the square of the length of the hypotenuse is equal to the sum of the squares of the lengths of the legs.

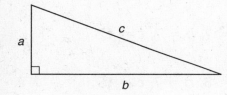

$c^2 = a^2 + b^2$, where $a$ and $b$ represent the lengths of the legs and $c$ represents the length of the hypotenuse.

If you know the lengths of two sides of a right triangle, you can use the Pythagorean Theorem to find the length of the third side. This is called **solving a right triangle**.

---

**EXAMPLE**    *Find the length of the hypotenuse of the right triangle.*

| | |
|---|---|
| $c^2 = a^2 + b^2$ | *Pythagorean Theorem* |
| $c^2 = 6^2 + 8^2$ | *Replace a with 6 and b with 8.* |
| $c^2 = 36 + 64$ | *Evaluate $6^2$ and $8^2$.* |
| $c^2 = 100$ | *Add 36 and 64.* |
| $\sqrt{c^2} = \sqrt{100}$ | *Take the square root of each side.* |
| $c = 10$ | *Simplify.* |

The length of the hypotenuse is 10 meters.

---

**EXERCISES** *In Exercises 1–6, find the area of each figure shown or described.*

**1.**

*c* ft
4 ft
3 ft

**2.**
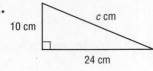
10 cm
*c* cm
24 cm

**3.**

5 in.
6 in.
*c* in.

**4.**
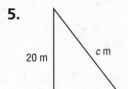
14 yd
3 yd
*c* yd

**5.**
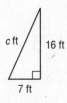
20 m
*c* m
15 m

**6.**
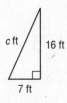
*c* ft
16 ft
7 ft

*If c is the measure of the hypotenuse, find each missing measure.*
*Round to the nearest tenth, if necessary.*

**7.** $a = 12$, $b = ?$, $c = 18$

**8.** $a = ?$, $b = 7$, $c = 12$

**9.** $a = ?$, $b = 24$, $c = 32$

**10.** $a = 8$, $b = ?$, $c = 15$

**11.** $a = 42$, $b = ?$, $c = 60$

**12.** $a = ?$, $b = 16$, $c = 20$

**APPLICATIONS**

**13.** Zach is working on a hat which involves a pattern made by fitting together pieces of fabric in the shape of right triangles. Each of the pieces of fabric has legs measuring 8 inches and 10 inches. Find the hypotenuse of each piece of fabric. Round to the nearest tenth.

**14.** Breanna is building a new house on a plot of land that is shaped like a right triangle. One of the legs of the plot measures 48 feet, and the hypotenuse measures 82 feet. Find the length of the other leg. Round to the nearest tenth.

# Three-Dimensional Figures

$C$ommon three-dimensional figures, or solids, are shown below.

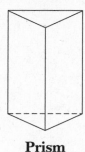

**Prism**

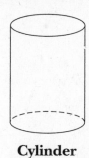

**Pyramid**

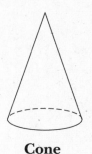

**Cylinder**

**Cone**

The flat surfaces of a prism or pyramid are called **faces**. The faces intersect to form **edges**. The edges intersect to form **vertices**.

At least two faces of a prism must be parallel and congruent polygons. These are called the **bases**. One face of a pyramid can be any polygon, also called the base. The bases of cylinders and cones are circles.

Prisms and pyramids are named by their bases. The prism at the right is a square prism.

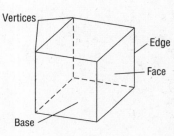

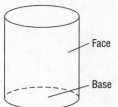

---

**EXAMPLES**   *Refer to the triangular prism.*

List the faces of the prism. Identify the bases.
Faces: △ABC, △DEF, ACFD, ABED, and BCFE
Bases: △ABC and △DEF

List the edges of the prism.
$\overline{AB}$, $\overline{BC}$, $\overline{AC}$, $\overline{AD}$, $\overline{BE}$, $\overline{CF}$, $\overline{DE}$, $\overline{EF}$, and $\overline{DF}$

List the vertices of the prism.
A, B, C, D, E, F

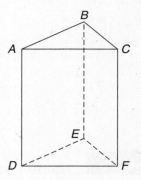

---

## EXERCISES  *Name each solid. Identify the bases.*

**1.**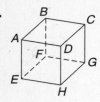

**2.**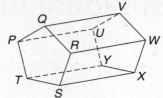

**3.**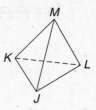

**4.**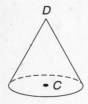

**5.**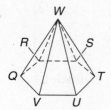

**6.**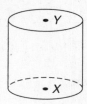

**7.**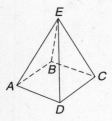

**8.**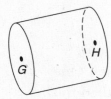

**9.**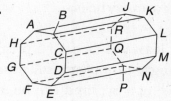

## Draw each solid.

**10.** hexagonal prism

**11.** pentagonal pyramid

**12.** cone

## APPLICATIONS  *Name two real-world examples of each solid.*

**13.** rectangular prism

**14.** square pyramid

**15.** cylinder

**16.** triangular pyramid

**17.** pentagonal prism

**18.** cone

**Name** _____ **Date** _____

# Customary Units of Measure

$C$ustomary units of length are inches, feet, yards, and miles.

Customary units of capacity are fluid ounces, cups, pints, quarts, and gallons.

Customary units of weight are ounces, pounds, and tons.

| Length |
|---|
| 1 foot (ft) = 12 inches (in.) |
| 1 yard (yd) = 36 inches or 3 feet |
| 1 mile (mi) = 5,280 feet or 1,760 yards |
| **Liquid Capacity** |
| 1 cup (c) = 8 fluid ounces (fl oz) |
| 1 pint (pt) = 2 cups |
| 1 quart (qt) = 2 pints |
| 1 gallon (gal) = 4 quarts |
| **Weight** |
| 1 pound (lb) = 16 ounces (oz) |
| 1 ton (T) = 2,000 pounds |

When changing from a smaller unit to a larger unit, divide.
When changing from a larger unit to a smaller unit, multiply.

**EXAMPLES**    *Change 29 inches to feet.*

29 in. = _____ ft     *You are changing from a smaller unit (in.) to a larger unit (ft), so divide.*

$29 \div 12 = 2\frac{5}{12}$     *Since 1 foot = 12 inches, divide by 12.*

$29 \text{ in.} = 2\frac{5}{12} \text{ ft}$

*Change 8 gallons to quarts.*

8 gal = _____ qt     *You are changing from a larger unit (gal) to a smaller unit (qt), so multiply.*

$8 \times 4 = 32$     *Since 1 gallon = 4 quarts, multiply by 4.*
8 gal = 32 qt

*Change 4 ounces to pounds.*

4 oz = _____ lb     *You are changing from a smaller unit (oz) to a larger unit (lb), so divide.*

$4 \div 16 = \frac{1}{4}$     *Since 1 pound = 16 ounces, divide by 16.*

$4 \text{ oz} = \frac{1}{4} \text{ lb}$

**EXERCISES**  *Complete.*

1. 5 ft = _____ in.

2. 36 in. = _____ ft

3. 9 ft = _____ yd

4. 7 yd = _____ ft

5. 5 mi = _____ ft

6. 10,560 ft = _____ mi

7. 55 in. = _____ ft

8. $1\frac{1}{2}$ yd = _____ in.

9. 10 ft = _____ yd

10. 3 c = _____ fl oz

11. 64 fl oz = _____ c

12. 36 c = _____ pt

13. 7 pt = _____ c

14. 8 gal = _____ qt

15. 100 qt = _____ gal

16. 9 pt = _____ qt

17. $3\frac{1}{2}$ gal = _____ qt

18. 3 gal = _____ c

19. 5 lb = _____ oz

20. 48 oz = _____ lb

21. 8 T = _____ lb

22. 12,000 lb = _____ T

23. 1,600 lb = _____ T

24. $3\frac{1}{2}$ T = _____ lb

**APPLICATIONS**  *Choose the best estimate.*

25. height of a giraffe          19 in.          19 ft          19 yd

26. length of a pencil           6 in.           6 ft           6 yd

27. lemonade in a pitcher        16 c            16 qt          16 gal

28. water in an aquarium         10 c            10 pt          10 gal

29. weight of a tennis ball      2 oz            2 lb           2 T

30. weight of a house cat        12 oz           12 lb          12 T

31. Lauren bought 4 gallons of punch for the party. If the glasses she plans to use for the party each hold one cup of liquid, how many glasses will she have?

32. Diego is 5 feet tall. John is 61 inches tall. Who is taller?

**Name** _____ **Date** _____

# Metric Units of Measure

Metric units of length are millimeters, centimeters, meters, and kilometers.

Metric units of capacity are milliliters, liters, and kiloliters.

Metric units of mass are milligrams (mg), grams (g), and kilograms (kg).

| Length |
|---|
| 1 centimeter (cm) = 10 millimeters (mm) |
| 1 meter (m) = 100 centimeters |
| 1 meter = 1,000 millimeters |
| 1 kilometer (km) = 1,000 meters |
| **Capacity** |
| 1 liter (L) = 1,000 milliliters (mL) |
| 1 kiloliter (kL) = 1,000 liters |
| **Mass** |
| 1 gram (g) = 1,000 milligrams (mg) |
| 1 kilogram (kg) = 1,000 grams |

When changing from a smaller unit to a larger unit, divide.
When changing from a larger unit to a smaller unit, multiply.

**EXAMPLES**  *Change 65 meters to centimeters.*

65 = 5 _____ cm      *You are changing from a larger unit (m)*
                      *to a smaller unit (cm), so multiply.*
65 × 100 = 6,500     *Since 1 meter = 100 centimeters, multiply by 100.*
65 m = 6,500 cm

*Change 500 milliliters to liters.*

500 mL = _____ L     *You are changing from a smaller unit (mL)*
                      *to a larger unit (L), so divide.*
500 ÷ 1,000 = 0.5    *Since 1 liter = 1,000 milliliters, divide by 1,000.*

*Change 4,500 grams to kilograms.*

4,500 g = _____ kg   *You are changing from a smaller unit (g)*
                      *to a larger unit (kg), so divide.*
4,500 ÷ 1,000 = 4.5  *Since 1 kilogram = 1,000 grams, divide by 1,000.*
4,500 g = 4.5 kg

## EXERCISES  *Complete.*

1. 55 mm = _____ cm
2. 71 cm = _____ mm
3. 7 km = _____ m

4. 750 m = _____ km
5. 210 cm = _____ m
6. 1.4 m = _____ cm

7. 5 m = _____ mm
8. 900 mm = _____ m
9. 3 km = _____ mm

10. 70 L = _____ mL
11. 9000 mL = _____ L
12. 0.6 kL = _____ L

13. 52 L = _____ kL
14. 70,000 mL = _____ kL
15. 90 mL = _____ kL

16. 16 kg = _____ g
17. 66 g = _____ kg
18. 1.2 g = _____ kg

## APPLICATIONS  *Choose the best estimate.*

19. length of a race — 5 cm — 5 m — 5 km
20. wingspan of an eagle — 2.4 cm — 2.4 m — 2.4 km
21. length of a computer disk — 90 mm — 90 cm — 90 m
22. capacity of can of soft drink — 355 mL — 355 L — 355 kL
23. capacity of a bathtub — 80 mL — 80 L — 80 kL
24. amount of vanilla in a cookie recipe — 3 mL — 3 L — 3 kL
25. mass of a nickel — 5 mg — 5 g — 5 kg
26. mass of an adult human — 60 mg — 60 g — 60 kg
27. mass of an apple — 0.2 mg — 0.2 g — 0.2 kg

28. The average shower uses 19 liters of water per minute. If you take a five-minute shower each day, how many kiloliters of water do you use in a 30-day month?

29. Soft drinks are sold in 2 liter containers. How many milliliters of soft drink are in one of these containers?

30. The mass of a collie is 33,000 grams, and the mass of a basset hound is 26 kilograms. Which dog is bigger?

Name _____ Date _____

# Converting Between Customary and Metric Units

You can use the following charts to convert between customary and metric units.

| Customary Unit | Metric Equivalent |
| --- | --- |
| 1 inch | 2.54 centimeters |
| 1 foot | 30.48 centimeters or 0.3048 meter |
| 1 yard | 0.914 meter |
| 1 mile | 1.609 kilometers |
| 1 ounce | 28.350 grams |
| 1 pound | 454 grams or 0.454 kilogram |
| 1 fluid ounce | 29.574 milliliters |
| 1 quart | 0.946 liter |
| 1 gallon | 3.785 liters |

| Metric Unit | Customary Equivalent |
| --- | --- |
| 1 centimeter | 0.394 inch |
| 1 meter | 3.281 feet or 1.093 yards |
| 1 kilometer | 0.621 mile |
| 1 gram | 0.035 ounce |
| 1 kilogram | 2.205 pounds |
| 1 milliliter | 0.034 fluid ounce |
| 1 liter | 1.057 quart or 0.264 gallon |

**Dimensional analysis** is a convenient method of converting between units of measure.

**EXAMPLES** *Complete.*

$5 \text{ mi} = \_\_\_\_\_ \text{ km}$

$5 \text{ mi} = \dfrac{5 \text{ mi}}{1} \times \dfrac{1.609 \text{ km}}{1 \text{ mi}}$   *Since 1 mi. is equivalent to 1.609 km, multiply 5 mi by 1.609 km per hour.*

$= \dfrac{5 \text{ mi}}{1} \times \dfrac{1.609 \text{ km}}{1 \text{ mi}}$   *Cancel common units in the numerator and denominator.*

$= 5 \times 1.609 \text{ km}$   *Simplify.*
$= 8.045 \text{ km}$   *Multiply.*

$40 \text{ kg} = \_\_\_\_\_ \text{ lb}$

$40 \text{ kg} = \dfrac{40 \text{ kg}}{1} \times \dfrac{2.205 \text{ lb}}{1 \text{ kg}}$   *1 kg = 2.205 lb*

$= \dfrac{40 \text{ kg}}{1} \times \dfrac{2.205 \text{ lb}}{1 \text{ kg}}$   *Cancel common units in the numerator and denominator.*

$= 40 \times 2.205 \text{ lb}$   *Simplify.*
$= 88.2 \text{ lb}$   *Multiply.*

**EXERCISES** *Complete. Round to the nearest thousandth if necessary.*

1. 6 in. = _____ cm

2. 7 qt = _____ L

3. 36 oz = _____ g

4. 2 gal = _____ L

5. 12 lb = _____ kg

6. 110 yd = _____ m

7. 4.25 m = _____ ft

8. 12.35 kg = _____ lb

9. 20.5 L = _____ gal

10. 245 mL = _____ fl oz

11. 228 km = _____ mi

12. 24 g = _____ oz

13. 63 in. = _____ m

14. 3.4 kg = _____ oz

15. 0.75 L = _____ fl oz

16. 500 mg = _____ oz

17. $2\frac{1}{4}$ c = _____ mL

18. 1,740 yd = _____ km

## APPLICATIONS

19. The speed limit in front of Central High School is 20 miles per hour. Express the speed limit in kilometers per hour.

20. At Lee's Farm Market, beans are on sale for $0.89 per pound. What is the cost per kilogram?

21. A cubic centimeter is equal to 1 milliliter. How many cubic centimeters does a 12-ounce glass hold?

22. If a house covers 1,700 square feet, how many square meters does it cover? Round to the nearest tenth.

23. What is the fuel economy in miles per gallon of a car that averages 12 kilometers per liter? Round to the nearest tenth.

**Name** _____ **Date** _____

# Units of Time

To add measures of time, add the seconds, add the minutes, and add the hours. Rename if necessary.

| Time |
|------|
| 1 minute (min) = 60 seconds (s) |
| 1 hour (h) = 60 minutes |

**EXAMPLE**   *Add 6 hours 35 minutes and 7 hours 45 minutes.*

$$6 \text{ h } 35 \text{ min}$$
$$+ 7 \text{ h } 45 \text{ min}$$
$$\overline{13 \text{ h } 80 \text{ min}} = 14 \text{ h } 20 \text{ min}$$

*Rename 80 min as 1 h 20 min. Then add to 13 h.*

To subtract measures of time, rename if necessary.
Then, subtract the seconds, subtract the minutes, and subtract the hours.

**EXAMPLE**   *Subtract 3 hours 18 minutes 55 seconds from 5 hours 20 minutes 10 seconds.*

$$5 \text{ h } 20 \text{ min } 10 \text{ s} = 5 \text{ h } 19 \text{ min } 70 \text{ s}$$
$$- 3 \text{ h } 18 \text{ min } 55 \text{ s} \quad - 3 \text{ h } 18 \text{ min } 55 \text{ s}$$
$$\overline{\phantom{- 3 \text{ h } 18 \text{ min } 55 \text{ s}}\, 2 \text{ h } 1 \text{ min } 15 \text{ s}}$$

*Rename 20 min 10 s as 19 min 70 s.*

**EXERCISES**   *Complete.*

1.  3 min 76 s = _____ min 16 s

2.  7 h 12 min = 6 h _____ min

3.  2 h 15 min 10 s = 2 h _____ min 70 s

4.  8 h 15 min 12 s = 7 h _____ min 12 s

5.  7 h 36 min = 6 h _____ min

6.  23 min 100 s = _____ min 40 s

7.  16 h 78 min = _____ h 18 min

8.  17 min 51 s = 16 min _____ s

9.  4 h 16 min 5 s = _____ h 76 min 5 s

10.  6 h 35 min 12 s = 5 h _____ min 72 s

*Add or subtract. Rename if necessary.*

11.        7 h 30 min    12.       35 min 28 s    13.      14 h 16 min
         − 5 h 18 min           + 15 min 22 s          + 12 h 52 min

14.    5 h 38 min   5 s    15.   4 h 45 min 17 s    16.    8 h 30 min
     − 2 h 20 min 15 s       + 7 h 23 min 18 s      − 2 h 10 min 15 s

17.       19 h 22 min    18.       42 min 12 s    19.        5 h 46 min
        − 8 h 44 min           − 33 min 42 s           + 5 h 33 min

20.      33 min 42 s    21.      17 h           22.       54 min   5 s
       + 11 min 22 s         − 8 h 37 min          − 44 min 23 s

## APPLICATIONS

23. It takes 2 hours 51 minutes to drive from New York City to Albany. It takes 3 hours 54 minutes to drive from Albany to Montreal. How long will it take to drive from New York City to Montreal going through Albany?

24. Miranda works after school from 4:30 P.M. to 8:15 P.M. How long does Miranda work? Give your answer in hours and minutes.

25. School starts at 8:45 A.M. It ends at 2:55 P.M. How long is the school day? Give your answer in hours and minutes.

26. The Anderson family left their house at 10:55 A.M. and arrived at their destination at 2:12 P.M. How long was their trip? Give your answer in hours and minutes.

27. The Wong family went on a two-week vacation. If they left on August 12, when will they return?

28. Shelandria went to see her doctor on April 17. Her doctor wants to see her again in three weeks. When should she return to the doctor's office?

**Name** _____ **Date** _____

# Precision and Significant Digits

$\mathsf{T}$he **precision** of a measurement is the exactness to which a measurement is made and is dependent upon the smallest unit of measure being used. **Significant digits** indicate the precision of the measurement.

---

**EXAMPLE**

*The cookie at the right has a diameter of 0.070 meter. Give the precision of this measurement. Then, give the measurements between which the actual length lies and the number of significant digits.*

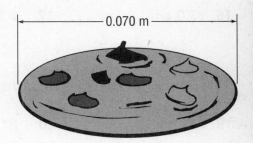

0.070 m

The precision of this measurement is the unit of measure, which is 0.001 meter.

The measurements between which the actual length lies are between 0.069 meter and 0.071 meter.

There are two significant digits. The zeros before the 7 are used only to show the place value of the decimal and are not counted as significant digits. However, the zero after the 7 is significant.

---

**EXERCISES** *Give the precision of each measure.*

**1.** 126 cm

**2.** 14 in.

**3.** 4.5 cm

**4.** 34 ft

**5.** 0.5 m

**6.** 19.04 m

**7.** 65 mm

**8.** 14 ft

**9.** 3.45 cm

---

*For each measure, give the measurements between which the actual length lies.*

**10.** 15 ft

**11.** 25 yd

**12.** 1.2 cm

**13.** 3.61 m

**14.** 79 in.

**15.** 453 mm

**16.** 17.0 cm

**17.** 0.004 m

**18.** 5.01 m

*Determine the number of significant digits in each measure.*

**19.** 142 cm

**20.** 81 in.

**21.** 10,233 mi

**22.** 0.022 cm

**23.** 4.05 m

**24.** 20.40 cm

**25.** 4 yd

**26.** 0.05 m

**27.** 505 ft

## APPLICATIONS

**28.** The sizes of men's hats are given as the diameter of a circle before the hat is made more oval. Hat sizes start at $6\frac{1}{4}$ and go by $\frac{1}{8}$-inch increments. How precise are the hat sizes?

**29.** Which would be the most precise measurement for a can of tomatoes: 2 pounds, 34 ounces, or 34.3 ounces?

**30.** Which is the more precise measurement for the height of the Statue of Liberty: 100 yards or 305 feet?

**31.** Noah and Blake both measure the same piece of copper tubing. Noah says the length is 22.8 centimeters. Blake says the length is 228.1 millimeters. Whose measurement is more precise?

**32.** Describe a situation where a measurement can be precise to the nearest mile.

**Name** _____ **Date** _____

# Perimeter and Area

The distance around a geometric figure is called **perimeter**.
The measure of the surface enclosed by a geometric figure is called **area**.

**EXAMPLE**

*Katsu wants to build a fence around a part of her backyard for her dog to play. She has 28 feet of fence. If the fence is to form a rectangle, what dimensions should Katsu use to give her dog the greatest area possible?*

| Dimensions | Perimeter | Area |
|:---:|:---:|:---:|
| 1 × 13 | 28 | 13 |
| 2 × 12 | 28 | 24 |
| 3 × 11 | 28 | 33 |
| 4 × 10 | 28 | 40 |
| 5 × 9 | 28 | 45 |
| 6 × 8 | 28 | 48 |
| 7 × 7 | 28 | 49 |

Notice that the perimeter stays 28 feet, but the area continues to increase. The greatest area with a perimeter of 28 feet is a square that is 7 feet by 7 feet.

**EXERCISES** *Find the perimeter and area of each figure.*

1.

2.
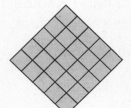

3.
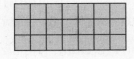

**4.** 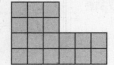  **5.** 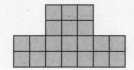  **6.**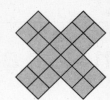

## APPLICATIONS

7. The figure at the right is the floor plan of a family room. The plan is drawn on grid paper, and each square of the grid represents one square foot.

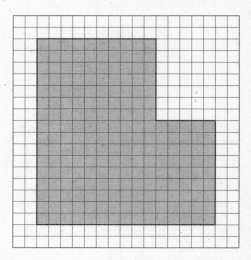

   a. What is the perimeter of the room?

   b. What is the area of the floor?

   c. The floor is going to be covered with square tiles. If each tile is 1 foot long, how many tiles will be needed?

   d. If each tile costs $3.25, how much will it cost to tile the floor?

8. A soccer field is 100 meters by 73 meters.

   a. The groundskeeper wants to outline the field. What is the perimeter of the field?

   b. What is the area of the field?

9. The Harris family wants to build a rectangular patio at the back of their house. They want the area of the patio to be at least 270 square feet. The length of the patio is to be 18 feet in length and the width can be no more than 16 feet.

   a. Is there room to build the size patio they want?

   b. What is the largest patio they can build?

   c. If they build a patio that is exactly 270 square feet, how wide should they make the patio?

**Name** _____ **Date** _____

# Area of Parallelograms, Rectangles, and Squares

Area is the number of square units needed to cover a surface.

|  | Rectangle | Square | Parallelogram |
|---|---|---|---|
| **Area Formula** | $A = \ell w$<br>$\ell$ = length and<br>$w$ = width | $A = s^2$<br>($s$ = side) | $A = bh$<br>$b$ = base and<br>$h$ = height |
| **Example** | 8 in.<br>14 in.<br>$A = \ell w$<br>$A = 14 \times 8$<br>$A = 112$<br>The area is 112 square inches. | 7 ft.<br>$A = s^2$<br>$A = 7^2$<br>$A = 49$<br>The area is 49 square feet. | 3 m<br>8 m<br>$A = bh$<br>$A = 8 \times 3$<br>$A = 24$<br>The area is 24 square meters. |

**EXERCISES** *Find the area of each figure.*

**1.**
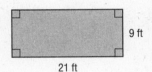
9 ft
21 ft

**2.**
15 cm
15 cm

**3.**
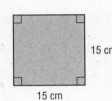
12 in.
16 in.

**4.**
20 m
20 m

**5.**
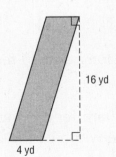
16 yd
4 yd

**6.**
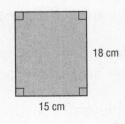
18 cm
15 cm

*Find the area of each figure.*

7. parallelogram
   base: 15 inches
   height: 11 inches

8. square
   side: 7.5 centimeters

9. rectangle
   length: 33 meters
   width: 21 meters

10. rectangle
    length: 16 feet
    width: 5 feet

11. parallelogram
    base: 3.5 yards
    height: 4.5 yards

12. square
    side: 44 millimeters

13. Find the area of a parallelogram with base 5.4 meters and height 3.2 meters.

14. What is the area of a rectangle with length 12.5 centimeters and width 6.5 centimeters?

15. What is the area of a square with side 0.65 meter?

16. What is the width of a rectangle if the length is 24 inches and the area is 360 square inches?

17. What is the length of the side of a square if the area is 16 square yards?

18. Find the height of a parallelogram if the base is 24 centimeters and area is 384 square centimeters.

## APPLICATIONS

19. What is the area of a regulation-size volleyball court with a length of 59 feet and width of 29.5 feet?

20. A high school basketball court is 84 feet long and 50 feet wide. What is the area of the court?

21. The area of a volleyball court is 1,800 square feet. If the length of the court is 60 feet, what is the width of the court?

**Name** _____ **Date** _____

# Area of Triangles and Trapezoids

A triangle is a polygon with three sides. A trapezoid is a quadrilateral with exactly one pair of parallel sides. The parallel sides are called bases.

|  | Triangle | Trapezoid |
|---|---|---|
| **Area Formula** | $A = \frac{1}{2}bh$<br>$b = base$ and<br>$h = $ height | $A = \frac{1}{2}h(a + b)$<br>$h = $ height,<br>$a = $ length of one base, and<br>$b = $ length to the other base |
| **Example** | 7 m<br>6 m<br><br>$A = \frac{1}{2}bh$<br>$A = \frac{1}{2} \times 6 \times 7$<br>$A = 21$<br>The area is 121 square meters. | 6 in.<br>5 in.<br>4 in.<br><br>$A = \frac{1}{2}h(a + b)$<br>$A = \frac{1}{2} \times 5(6 + 4)$<br>$A = 25$<br>The area is 25 square inches. |

**EXERCISES** *Find the area of each figure.*

**1.**

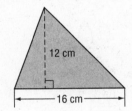

12 cm
16 cm

**2.**

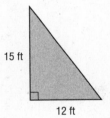

15 ft
12 ft

**3.**

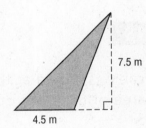

7.5 m
4.5 m

**4.**

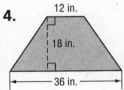

12 in.
18 in.
36 in.

**5.**

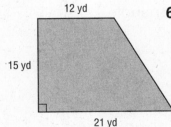

12 yd
15 yd
21 yd

**6.**

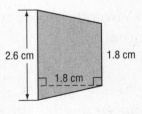

2.6 cm
1.8 cm
1.8 cm

**Find the area of each figure.**

7. triangle
   base: 7 centimeters
   height: 12 centimeters

8. triangle
   base: 9 meters
   height: 7 meters

9. trapezoid
   bases: 7 inches, 9 inches
   height: 5 inches

10. trapezoid
    bases: 6 m, 9 m
    height: 11 meters

11. triangle
    base: 15 yards
    height: 9 yards

12. trapezoid
    bases: 11 inches, 8 inches
    height: 13 inches

13. Find the area of a triangle with a base of 44.5 centimeters and a height of 36 centimeters.

14. Find the area of a trapezoid with bases of 3.7 meters and 5.8 meters and height of 4.8 meters.

15. What is the height of a triangle with a base of 7 feet and an area of 42 square feet?

16. A trapezoid has an area of 660 square yards. If the lengths of the bases are 19 yards and 25 yards, what is the height?

## APPLICATIONS

17. An A-frame cabin is shown at the right. What is the area of the front of the cabin?

18. The shape of Virginia is close to a triangle. Estimate its area.

19. The maintenance crew for the city parks wants to plant grass in an area shown in the diagram at the right. If a bag of grass seed covers 2,000 square feet, how many bags will they need?

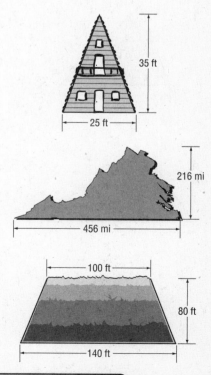

35 ft

25 ft

216 mi

456 mi

100 ft

80 ft

140 ft

Name _____    Date _____

# Area of Irregular Shapes

$O$ne way to estimate the area of an irregular figure is to find the mean of the inner measure and the outer measure of the figure.

---

**EXAMPLE**    *Estimate the area of the figure.*

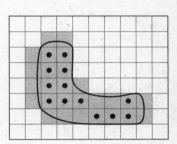

inner measure (marked with dots): 13
outer measure (all shaded squares): 14
mean: $\frac{13 + 14}{2} = 13.5$

An estimate of the area of the figure is 13.5 square units.

---

$S$ometimes an irregular shape can be separated into two or more shapes whose areas you know how to find.

---

**EXAMPLE**    *Find the area of the figure.*

The figure can be separated into a
triangle and a trapezoid.

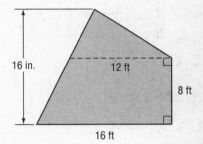

**Area of Triangle**

The base of the triangle is 12 feet,
and the height is 14 − 8 or 6 feet.

$A = \frac{1}{2}bh$

$A = \frac{1}{2} \times 12 \times 6$

$A = 36$

**Area of Trapezoid**

The bases of the trapezoid are 12 feet
and 16 feet, and the height is 8 feet.

$A = \frac{1}{2}h(a + b)$

$A = \frac{1}{2} \times 8(12 + 16)$

$A = 112$

The area of the figure is 36 + 112 or 148 square feet.

---

**EXERCISES** *Estimate the area of each figure.*

**1.**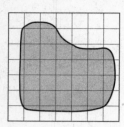

**2.**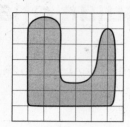

**3.**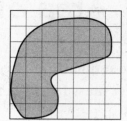

**4.**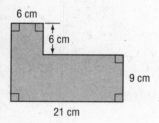

6 cm
6 cm
9 cm
21 cm

**5.**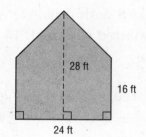

28 ft
16 ft
24 ft

**6.**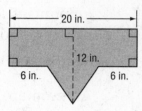

20 in.
12 in.
6 in.  6 in.

**7.**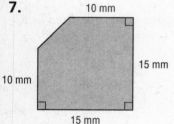

10 mm
15 mm
10 mm
10 mm
15 mm

**8.**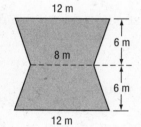

12 m
6 m
8 m
6 m
12 m

**9.**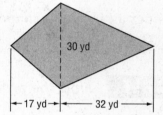

30 yd
17 yd  32 yd

**APPLICATIONS**

**10.** A diagram of an L-shaped room is shown at the right. If carpet costs $3.50 per square foot, find the cost to carpet the room.

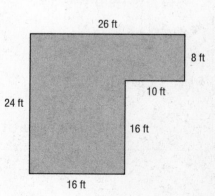

26 ft
8 ft
10 ft
24 ft
16 ft
16 ft

**11.** Draw an outline of your foot on grid paper. Then, estimate the area of your foot.

**Name** _____ **Date** _____

# Circumference and Area of Circles

The parts of a circle are illustrated at the right. Notice that the radius is one-half of the diameter.

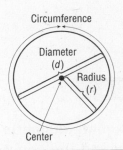

|  | Circumference | Area |
|---|---|---|
| **Formula** | $C = \pi d$ or $C = 2\pi r$ <br> ($d$ = diameter and $r$ = radius) | $A = \pi r^2$ <br> ($r$ = radius) |
| **Example** | 7 in. <br><br> $C = 2\pi r$ <br> $C = 2\pi \times 7$ <br> $C \approx 44.0$  Use a calculator <br> The circumference is about 44.0 inches. | 18 cm <br><br> The radius is half of 18 or 9 centimeters <br> $A = \pi r^2$ <br> $A = \pi \times 9^2$ <br> $A \approx 254.5$  Use a calculator. <br> The area is about 254.5 square centimeters. |

**EXERCISES** *Find the circumference and area of each circle. Round to the nearest tenth.*

1.
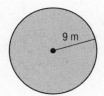
9 m

2.
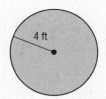
4 ft

3.
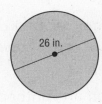
26 in.

4.

34 mm

5.
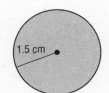
1.5 cm

6.
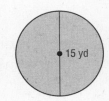
15 yd

**7.**

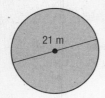

21 m

**8.**

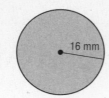

16 mm

**9.**

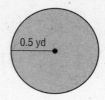

0.5 yd

## APPLICATIONS

10. The approximate diameter of Earth is 3,960 miles. What is the distance around the equator?

11. A circular garden has a diameter of 28 feet. The garden is to be covered with peat moss. If each bag of peat moss covers 160 square feet, how many bags of peat moss will be needed?

12. What is the area of a pizza with a diameter of 14 inches?

13. The stage of a theater is a semicircle. If the radius of the stage is 32 feet, what is the area of the stage?

14. A Ferris wheel has a diameter of 212 feet. How far will a passenger travel in one revolution of the wheel?

15. A water sprinkler produces a spray that goes out 25 feet. If it sprays the water in a circular pattern, what is the area of the lawn that it waters?

16. The diameter of a bicycle wheel is 26 inches. How many feet will the bicycle travel if the wheel turns 20 times?

17. The Roman Pantheon is 142 feet in diameter. What is the distance around the Pantheon?

18. The diameter of a dime is 17.9 millimeters. What is the area of one side of the coin?

**Name** _____  **Date** _____

# Nets and Solids

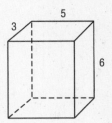

$A$ **net** is the shape that is formed by "unfolding" a three-dimensional figure. A net shows all the faces that make up the surface area of the figure.

**EXAMPLE**

*Draw a net for a rectangular prism that has length 5 units, width 3 units, and height 6 units.*

The net will be made up of three sets of congruent shapes:

1) the top and bottom of the prism— rectangles 3 units by 5 units

2) the two sides of the prism— rectangles 3 units by 6 units

3) the front and back of the prism— rectangles 5 units by 6 units.

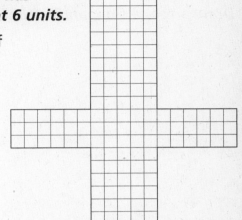

$T$hree-dimensional figures can also be portrayed by drawing their top, side, and front views separately.

**EXAMPLE**

*Draw a top, a side, and a front view for the pyramid*

If you look at the figure from directly above, you would see the square base.

Looking at the figure from the sides, front, and the rear you would see a triangle.

*Draw a net for each figure.*

**1.**

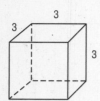

**2.**

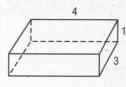

*Draw a top, a side, and a front view of each figure.*

**3.**

**4.**

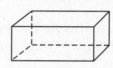

*Make a perspective drawing of each figure by using the top, side, and front views as shown.*

**5.**

**6.**

**APPLICATIONS**

7. Javier finds a container of oatmeal that is in the shape of a cylinder. Draw a top, a side, and a front view of the container.

8. Victoria has borrowed a block from her little sister's building block collection. The top, side, and front views of the block are given. Draw the block.

**SKILL 64**

**Name** _____ **Date** _____

# Volume of Rectangular Prisms and Cylinders

The volume of an object is the amount of space a solid contains. Volume is measured in cubic units. The volume of a prism or a cylinder can be found by multiplying the area of the base times the height.

$$V = Bh$$

For a rectangular prism, the area of the base equals the length times the width. For cylinders, the area of the base equals pi times the radius squared.

| | Rectangular Prism | Cylinder |
|---|---|---|
| **Volume Formula** | $V = \ell wh$ <br> ($\ell$ = length, $w$ = width, and $h$ = height) | $V = \pi r^2 h$ <br> ($r$ = radius and $h$ = height) |
| **Example** | $V = \ell wh$ <br> $V = 7 \times 5 \times 3$ <br> $V = 105$ <br> The volume is 105 cubic meters. | $V = \pi r^2 h$ <br> $V = \pi \times 10^2 \times 8$ <br> $V \approx 2{,}513.3$  Use a calculator. <br> The volume is about 2,513.3 cubic inches. |

**EXERCISES**  *Find the volume of each rectangular prism.*

1.
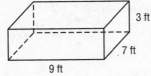
3 ft
7 ft
9 ft

2.
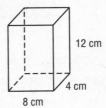
12 cm
4 cm
8 cm

3.
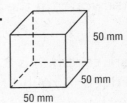
50 mm
50 mm
50 mm

4.
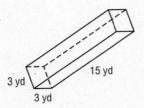
15 yd
3 yd
3 yd

5.
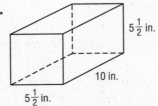
$5\frac{1}{2}$ in.
10 in.
$5\frac{1}{2}$ in.

6.
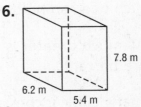
7.8 m
6.2 m
5.4 m

**Find the volume of each cylinder. Round to the nearest tenth.**

**7.**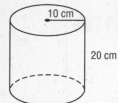
10 cm
20 cm

**8.**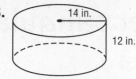
14 in.
12 in.

**9.**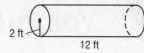
2 ft
12 ft

**10.**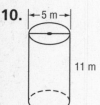
5 m
11 m

**11.**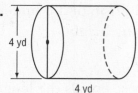
4 yd
4 yd

**12.**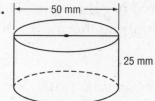
50 mm
25 mm

## APPLICATIONS

**13.** Mariah is making cylindrical candles. The candles she plans to make have a diameter of 3 inches and a height of 8 inches. If she has 200 cubic inches of wax, how many candles can Mariah make.

**14.** A fish tank is shown at the right.

  **a.** What is the volume of the tank?

  **b.** If the tank is filled to the height of 50 centimeters, what is the volume of the water in the tank?

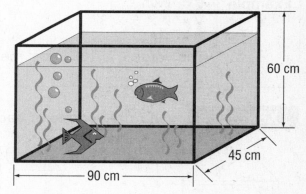
60 cm
45 cm
90 cm

**15.** A square cake pan is 8 inches long on each side. It is 2 inches deep. A round cake pan has a diameter of 8 inches. It is 2 inches deep.

  **a.** Which pan can hold more cake batter?

  **b.** How much greater is the volume of the pan that holds more batter?

**16.** How many cubic inches are in a cube that is 1 foot on each side?

**Name** _____     **Date** _____

# Surface Area of Rectangular Prisms and Cylinders

The sum of the areas of all the surfaces of a solid is called the **surface area**.

The surface area of a rectangular prism is the sum of the areas of each of its six faces.

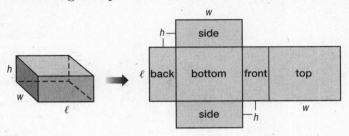

top and bottom: $(\ell \times w) + (\ell \times w) = 2\ell w$
front and back: $(\ell \times h) + (\ell \times h) = 2\ell h$
two sides: $(w \times h) + (w \times h) = 2wh$
surface area $= 2\ell w + 2\ell h + 2wh$

The surface area of a cylinder is the sum of the areas of the two bases and the curved surface.

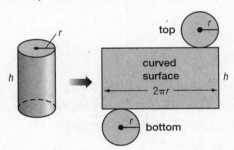

top and bottom: $\pi r^2 + \pi r^2 = 2\pi r^2$
curved surface: $(2\pi r) \times h = 2\pi rh$
surface area $= 2\pi r^2 + 2\pi rh$

---

**EXAMPLES**   *Find the surface area of each solid.*

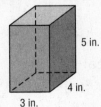

5 in.
4 in.
3 in.

surface area $= 2\ell w + 2\ell h + 2wh$
surface area $= 2 \times 3 \times 4 + 2 \times 3 \times 5 + 2 \times 4 \times 5$
surface area $= 94$
The surface area is 94 square inches.

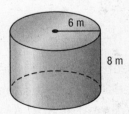

6 m
8 m

surface area $= 2\pi r^2 + 2\pi rh$
surface area $= 2\pi r^2 \times 6^2 + 2\pi \times 6 \times 8$
surface area $\approx 527.8$
The surface area is about 527.8 square meters.

---

**1.**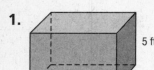
5 ft
3 ft
8 ft

**2.**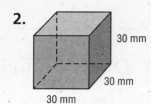
30 mm
30 mm
30 mm

**3.**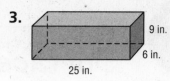
9 in.
6 in.
25 in.

**4.**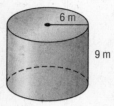
6 m
9 m

**5.**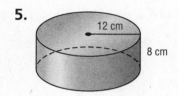
12 cm
8 cm

**6.**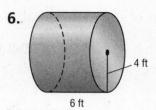
4 ft
6 ft

**7.**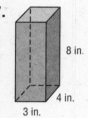
8 in.
4 in.
3 in.

**8.**
6 cm
15 cm

**9.**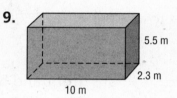
5.5 m
2.3 m
10 m

## APPLICATIONS

**10.** A box company is making rectangular boxes that are 10 centimeters by 8 centimeters by 5 centimeters. How many of these boxes can the company make using 200,000 square centimeters of cardboard?

**11.** The two boxes have about the same volume. Which box takes less material to manufacture?

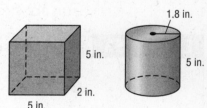
1.8 in.
5 in.
5 in.
2 in.
5 in.

**12.** A wheel of cheese is sealed in a wax covering. The wheel of cheese is in the shape of a cylinder that has a diameter of 25 centimeters and a height of 20 centimeters. What is the surface area of the cheese that needs to be covered in wax.

# Mean, Median, and Mode

You can analyze a set of data by using three measures of central tendency: **mean**, **median**, and **mode.**

**EXAMPLE**

*Tim Duncan, 2003's Most Valuable Player in the National Basketball Association, helped the San Antonio Spurs win the NBA Championship. In winning the six games of the series, Duncan scored 32, 19, 21, 23, 29, and 21 points. Find the mean, median, and mode of his scores.*

**Mean:** $\dfrac{32 + 19 + 21 + 23 + 29 + 21}{6} \approx 24.167$

The mean is about 24 points.

**Median:** 19, 21, 21, 23, 29, 32

$\dfrac{21 + 23}{2} = 22$

The median is 22 points.

**Mode:** The mode is 21 since it is the number that appears the most times.

**EXERCISES** *Find the mean, median, and mode for each set of data.*

1. 2, 3, 7, 8, 10, 3, 1, 7, 5

2. 17, 18, 20, 13, 23, 37, 20, 16

3. 4.8, 6.4, 7.2, 4.5, 2.3, 6.0, 3.5

4. 82, 77, 82, 76, 79, 78, 81, 86

5. 40, 42, 41, 43, 41, 40, 40, 42, 43

6. $7.50, $7.00, $8.50, $7.50, $4.50, $6.50, $8.00, $6.00, $4.50

**7.** 1.78, 1.45, 1.33, 1.72, 1.94, 1.73, 1.14

**8.** 3, −3, 1, 4, 5, 0, −4, −1, 2, −1

**9.** 90%, 98%, 96%, 85%, 91%, 90%, 88%, 87%, 88%, 90%

**10.** 5.8 cm, 8.9 cm, 8.8 cm, 8.6 cm, 8.8 cm, 8.8 cm, 8.9 cm

**11.** $50,000, $37,500, $43,900, $76,900, $46,000, $48,580

**12.** 29.1°F, 33.9°F, 38.2°F, 46.5°F, 55.4°F, 62.0°F, 63.6°F, 62.3°F, 56.1°F, 47.2°F, 37.3°F, 32.0°F

**APPLICATIONS**   *The data at the right shows the ages of U.S. Presidents from 1900–2004 at the time of their inaugurations. Use this data to answer Exercises 13–16.*

| 58 | 42 | 51 | 56 | 55 |
|----|----|----|----|----|
| 51 | 54 | 51 | 60 | 62 |
| 43 | 55 | 56 | 61 | 52 |
| 69 | 64 | 46 | 54 |    |

**13.** What is the mode of the data?

**14.** What is the mean of the data?

**15.** What is the median of the data?

**16.** If the age of each President was 1 year older, would it change
**a.** the mean? Why or why not?

**b.** the median? Why or why not?

**c.** the mode? Why or why not?

Name _____  Date _____

# Frequency Tables

There are four frequency values that can be considered.

**absolute frequency**: the frequency for an individual interval
**relative frequency**: the ratio of an interval's absolute frequency to the total number of elements
**cumulative frequency**: the sum of the absolute frequency of an interval and all previous absolute frequencies
**relative cumulative frequency**: the ratio of an interval's cumulative frequency to the total number of elements

**EXAMPLE**  *The base prices of new cars from one manufacturer are listed below.*

| $24,625 | $16,200 | $22,225 | $40,450 | $35,050 | $33,565 | $44,535 |
| $22,075 | $24,370 | $20,465 | $9,995 | $21,560 | $25,330 | $24,695 |
| $28,105 | $21,630 | $22,145 | $41,995 | $28,905 | $30,655 | $10,700 |
| $18,995 | $20,060 | $37,900 | $22,080 | | | |

*Organize this information in a frequency table. Determine the absolute frequencies, relative frequencies, cumulative frequencies and relative cumulative frequencies for the data. Then find the range of the data.*

Use intervals of 10,000 to organize the data.

| New Car Prices | | | | | |
|---|---|---|---|---|---|
| Interval | Tally | Absolute Frequency | Relative Frequency | Cumulative Frequency | Relative Cumulative Frequency |
| $0–$9,999 | \| | 1 | $\frac{1}{25} = 0.04$ | 1 | $\frac{1}{25} = 0.04$ |
| $10,000–$19,999 | \|\|\| | 3 | $\frac{3}{25} = 0.12$ | 4 | $\frac{4}{25} = 0.16$ |
| $20,000–$29,999 | \|\|\|\| \|\|\|\| \|\|\|\| | 14 | $\frac{14}{25} = 0.56$ | 18 | $\frac{18}{25} = 0.72$ |
| $30,000–$39,999 | \|\|\|\| | 4 | $\frac{4}{25} = 0.16$ | 22 | $\frac{22}{25} = 0.88$ |
| $40,000–$49,999 | \|\|\| | 3 | $\frac{3}{25} = 0.12$ | 25 | $\frac{25}{25} = 1.00$ |

To determine the range, find the difference between the highest price and lowest price.
$44,535 − $9,995 = $34,540

*Organize the information in a frequency table. Determine the absolute frequencies, relative frequencies, cumulative frequencies, and relative cumulative frequencies for the data.*

1. Test scores of students in a classroom.
   94, 81, 85, 59, 83, 73, 75, 96, 72, 87, 77, 88, 90, 65, 71, 82, 86, 89, 68, 96

*Use a frequency table to determine the absolute frequencies, relative frequencies, cumulative frequencies, and relative cumulative frequencies for the data.*

2. World Series champions from 1990–2003.

| | | |
|---|---|---|
| 1990 Cincinnati Reds | 1991 Minnesota Twins | 1992 Toronto Blue Jays |
| 1993 Toronto Blue Jays | 1994 *Not Held* | 1995 Atlanta Braves |
| 1996 New York Yankees | 1997 Florida Marlins | 1998 New York Yankees |
| 1999 New York Yankees | 2000 New York Yankees | 2001 Arizona Diamondbacks |
| 2002 Anaheim Angels | 2003 Florida Marlins | |

**Name** _____ **Date** _____

# Line Plots

The table lists the states that had a change in the number of U.S. Representatives from 1990-2000.

| State | AZ | CA | CO | CT | FL | GA | IL | IN | MI | MS | NV | NY | NC | OH | OK | PA | TX | WI |
|---|---|---|---|---|---|---|---|---|---|---|---|---|---|---|---|---|---|---|
| Change in Representation | +2 | +1 | +1 | −1 | +2 | +2 | −1 | −1 | −1 | −1 | +1 | −2 | +1 | −1 | −1 | −2 | +2 | −1 |

**EXAMPLE** *Organize this information using a line plot.*

The lowest change is −2, and the highest change is +2.
Draw a number line that includes the numbers −2 to +2.
Place an X above the number line to represent each state.

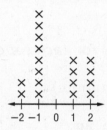

**EXERCISES** *Make a line plot for each set of data.*

**1.** 11, 11, 17, 14, 15, 14, 11, 13, 12, 11, 11, 15

**2.** 31, 27, 36, 31, 49, 38, 32, 46, 31, 47, 43, 25

**3.** 100, 90, 75, 85, 65, 100, 85, 80, 90, 90, 75, 70

**4.** 2.9, 2.0, 2.4, 2.8, 2.5, 2.3, 2.2, 2.9, 2.5, 2.8, 2.0, 2.8

**5.** −10, 8, 10, −7, −10, 3, 4, −6, −2, 10, −7, −6, 4, −9, −3

**6.** 520, 760, 720, 580, 620, 740, 560, 510, 670, 500, 600, 700, 770, 700

**7.** 0.12, 0.19, 0.15, 0.11, 0.14, 0.11, 0.14, 0.17, 0.16,
0.14, 0.13, 0.14, 0.13, 0.16, 0.15, 0.13

**8.** 2,000, 2,250, 2,200, 2,150, 2,050, 2,500, 2,550, 2,300,
2,350, 2,550 2,250, 2,350, 2,300, 2,000, 2,500, 2,300,
2,200, 2,550, 2,100, 2,250

**APPLICATIONS**   *Make a line plot for each set of data.*

**9.** The resting pulse rates (beats per minute)
of 11 athletes are listed below.
62, 59, 60, 61, 57, 48, 59, 64, 58, 60, 49

**10.** The costs for hair styling at various shops are listed below.
$25.00, $27.00, $32.00, $22.00, $43.00, $28.00, $18.00, $25.00
$24.00, $25.00, $27.00, $30.00, $24.00, $22.00, $30.00, $20.00

**11.** Ms. Madson asked her students what time they have dinner
to the nearest half hour. The results are listed below.
6:30, 7:30, 7:00, 6:30, 6:00, 7:00,
5:30, 7:00, 6:30, 6:00, 6:30, 7:00,
6:30, 6:30, 6:30, 7:00, 7:00, 6:30

**12.** Ask your classmates to rate a certain song from 1 to 10 with
10 being the best. Write down their responses and organize
the information in a line plot.

Name _____    Date _____

# Stem-and-Leaf Plots

A stem-and-leaf plot is one way to organize a list of numbers. The **stems** represent the greatest place value of all the numbers. The **leaves** represent the next place value.

---

**EXAMPLE**   *Make a stem-and-leaf plot for the following data.*

19, 30, 27, 34, 23, 34, 39, 48, 15, 47, 25
31, 46, 43, 36, 27, 35, 33, 21, 37, 27, 42

The stem will be the tens place and the leaves will be the ones place.

```
1 | 5 9
2 | 1 3 5 7 7 7
3 | 0 1 3 4 4 5 6 7 9
4 | 2 3 6 7 8
```

4│2 means 42.

---

**EXERCISES**   *Make a stem-and-leaf plot for each set of data.*

**1.** 100, 62, 85, 72, 99, 87, 87, 93, 77,
86, 96, 79, 100, 86, 94, 68, 75, 90,
88, 99, 87, 73, 66, 89, 67, 70, 93

**2.** 1.6, 2.3, 1.3, 2.2, 0.9, 1.1, 2.6, 1.6, 3.5, 4.2,
2.9, 2.4, 3.8, 1.4, 0.6, 1.7, 1.8, 1.2, 0.7, 1.9,
2.6, 1.7, 2.5, 1.5, 0.7, 2.4, 1.5, 1.8, 3.3, 2.5

**3.** 710, 460, 700, 710, 470, 520, 520, 820,
370, 670, 850, 450, 410, 560, 770, 350,
440, 340, 520, 810, 580, 790, 780, 520

**4.** 1, 24, 17, 18, 4, 38, 34, 42, 12, 48, 19, 30, 25,
44, 37, 37, 41, 23, 24, 2, 32, 33, 49, 40, 51, 8,
6, 21, 49, 41, 48, 4, 52, 15, 49, 22, 23, 11, 17

**5.** 10.1, 12.8, 13.3, 9.8, 12.0, 10.0, 13.6, 9.7, 10.2, 12.3, 13.7, 8.1, 10.4, 13.3, 11.0, 8.4, 11.6, 8.7, 8.4, 13.3, 13.6, 10.4, 10.1, 9.9

**6.** 0.10, 0.11, 0.58, 0.10, 0.25, 0.49, 0.24, 0.46, 0.16, 0.13, 0.21, 0.34, 0.16, 0.29, 0.28, 0.13, 0.14, 0.14, 0.20, 0.18, 0.18 0.21, 0.13, 0.22, 0.11, 0.23, 0.12

**APPLICATIONS** *Make a stem-and-leaf plot for each set of data.*

**7.** The high temperatures (°F) for a city for the month of January are listed below.
50, 57, 60, 46, 40, 29, 25, 30, 26, 20, 35, 38, 34, 39, 34, 22, 33, 34, 22, 21, 32, 33, 18, 20, 19, 37, 39, 22, 21, 12, 9

**8.** The students in Mrs. Palmer's math class received the following grades on a recent test.
94, 72, 68, 78, 74, 95, 68, 69, 67, 87, 83, 79, 99, 70, 82, 59
69, 80, 86, 87, 82, 100, 68, 63, 80, 91, 69, 82, 96, 89, 61

**9.** Find the length of your classmates' shoes in centimeters. Record the data and make a stem-and-leaf plot.

**Name** _____ **Date** _____

# Box-and-Whisker Plots

A **box-and-whisker plot** separates a set of data into four parts using the median and quartiles. A **box** is drawn around the quartile values, and **whiskers** extend from each quartile to the extreme data points.

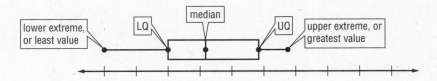

| **EXAMPLE** | *Draw a box-and-whisker plot for the set of test scores given below.* |
|---|---|

83   92   75   96   89   70   62   85   94   88   92

Step 1    Find the least and greatest numbers. Then draw a number line that covers the range of the data.
The lowest data value is 62 and the highest is 96. Draw a number line ranging from 60 to 100 with increments of 5.

Step 2    Find the median, the extremes, and the upper and lower quartiles. Mark these points above the number line.

The median for the data is 88. The extremes are 62 and 96. The upper quartile is 92 and the lower quartile is 75.

Step 3    Draw a box and the whiskers.

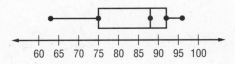

| **EXERCISES** | *Draw a box-and-whisker plot for each set of data.* |
|---|---|

1.   18, 24, 21, 19, 32, 22,
     45, 21, 18, 28, 30

2.   132, 118, 145, 107, 125, 130,
     105, 114, 122, 138, 117

**3.** 75, 88, 100, 92, 68, 73, 95, 84,
70, 85, 90, 66, 78, 89, 95

**4.** 6.8, 7.7, 8.3, 5.4, 6.9, 7.0,
9.1, 8.2, 7.1, 6.3, 5.5

**5.** 38, 42, 27, 19, 35, 40,
31, 24, 45, 37, 41

**6.** 5, 9, 7, 6, 5, 8, 4, 9,
7, 7, 6, 5, 8, 9, 6

**7.** $89, $74, $62, $83, $94, $66,
$80, $73, $88, $91, $70

**8.** 18, 21, 19, 18, 20,
22, 19, 24, 23

## APPLICATIONS

**9.** The following data gives the total amount of
snowfall (in inches) for a community in Ohio, over
the past nine winters. Draw a box-and-whisker
plot for the data.

24.3, 18.6, 21.9, 34.1, 17.4, 25.5, 31.3, 22.7, 24.6

**10.** David is in charge of counting the money
collected at his school each day for the annual
fund-raiser. The data below shows the amounts
collected each day during the past two weeks.
Draw a box-and-whisker plot for the data.

$12, $23, $18, $15, $9, $25, $14, $11, $21, $16

**Name** _____ **Date** _____

# Line Graphs

A **line graph** usually used to show the change and direction of change over time. All line graphs should have a graph title, a vertical-axis label, and a horizontal-axis label.

**EXAMPLE**  *Make a line graph for the data about the value of a company's stock over a one-week period.*

| SuperCola Stock | |
|:---:|:---:|
| **Day** | **Value ($)** |
| Monday | 32.25 |
| Tuesday | 34.00 |
| Wednesday | 36.50 |
| Thursday | 32.00 |
| Friday | 31.50 |

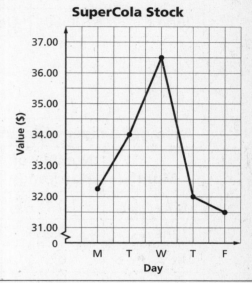

**EXERCISES**  *Make a line graph for each set of data.*

1.

| Number of Votes Expected | |
|:---:|:---:|
| **Date** | **Number of Votes** |
| 1/15 | 7 |
| 1/31 | 10 |
| 2/15 | 12 |
| 2/28 | 16 |
| 3/15 | 18 |
| 3/31 | 16 |
| 4/15 | 10 |
| 4/30 | 15 |
| 5/15 | 23 |

**2.**

| Ramerez Family Electric Bill | | | |
|---|---|---|---|
| Month | Amount | Month | Amount |
| January | $88.50 | July | $125.50 |
| February | $87.70 | August | $132.10 |
| March | $95.00 | September | $103.20 |
| April | $93.00 | October | $86.80 |
| May | $96.00 | November | $84.20 |
| June | $100.60 | December | $98.70 |

**3.**

| Average Number of Atlantic Hurricanes, 1950–2001 | | | |
|---|---|---|---|
| Month | Number of Hurricanes | Month | Number of Hurricanes |
| May | 0.1 | September | 2.75 |
| June | 0.25 | October | 1.0 |
| July | 0.5 | November | 0.25 |
| August | 1.75 | December | 0.1 |

**4.**

| Traffic at 4th Ave. and 10th St. | |
|---|---|
| Day | Number of Vehicles |
| Sunday | 62,000 |
| Monday | 72,000 |
| Tuesday | 80,500 |
| Wednesday | 76,500 |
| Thursday | 84,000 |
| Friday | 86,500 |
| Saturday | 68,000 |

**Name** _____ **Date** _____

# Bar Graphs and Histograms

**D**isplaying data in graphs makes it easier to visualize the data. A **bar graph** is used to display individual frequencies.

**EXAMPLE**   *The table shows the pizza sales at Paul's Pizza for one week. Construct a bar graph to show the data.*

| Paul's Pizza Weekly Sales | |
|---|---|
| Personal | 40 |
| Small | 80 |
| Medium | 125 |
| Large | 85 |
| Extra Large | 50 |

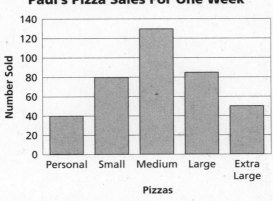

**Paul's Pizza Sales For One Week**

**A** **histogram** uses bars to display numerical data that have been organized into equal intervals.

**EXAMPLE**   *The table shows the percent of people in several age groups who are not covered by health insurance. Make a histogram of the data.*

| Age | Percent |
|---|---|
| 18–24 | 28.9% |
| 25–34 | 20.9% |
| 35–44 | 15.5% |
| 45–54 | 14.0% |
| 55–64 | 12.9% |

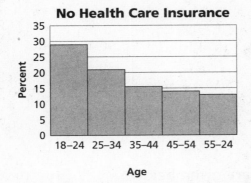

**No Health Care Insurance**

**APPLICATIONS** *Make a bar graph for each set of data.*

1.

**Preference for Colas**

| Brand | Number of Students |
|-------|--------------------|
| A | 64 |
| B | 72 |
| C | 18 |
| D | 40 |

2.

**Highest Mountains (feet above sea level)**

| Continent | Elevation |
|-----------|-----------|
| Africa | 19,340 |
| Antarctica | 16,864 |
| Asia | 29,035 |
| Australia | 16,500 |
| Europe | 15,771 |
| North America | 20,320 |
| South America | 22,834 |

*Make a histogram for each set of data.*

3.

**Time Using Internet (hours per day)**

| 0–1 | 20 |
|-----|----|
| 2–3 | 18 |
| 4–5 | 2 |
| 6–7 | 1 |

4. The test scores of a recent math test are given below.
   89, 93, 80, 79, 92, 96, 75, 84, 99, 76, 92,
   85, 87, 98, 97, 82, 85, 98, 81, 89, 81, 88

# Circle Graphs

A **circle graph** is used to show how parts of data are related to the whole set of data.

**EXAMPLE**   *The areas of Earth's oceans are listed below. Make a circle graph to show the area of Earth's oceans.*

| Ocean | Arctic | Atlantic | Indian | Pacific | Southern |
|---|---|---|---|---|---|
| Total ocean surface area (%) | 4.2 | 22.9 | 20.4 | 46.4 | 6.0 |

To make a circle graph, first find the number of degrees that correspond to each ocean. Use a calculator to round to the nearest degree.

Arctic:      4.2% of 360° ≈ 15°
Atlantic:    22.9% of 360° ≈ 82°
Indian:      20.4% of 360° ≈ 73°
Pacific:     46.4% of 360° ≈ 167°
Southern:    6.0% of 360° ≈ 22°

Use a compass and a protractor to draw the circle graph.

Note that the sum of the degrees is not 360° because of rounding.

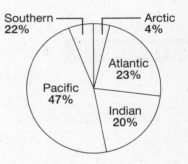

**Surface Area of Oceans**

**EXERCISES**   *Make a circle graph to show the data in each table.*

1.

| Favorite Music | |
|---|---|
| Country | 12% |
| Dance | 8% |
| Jazz | 10% |
| Pop | 23% |
| Rap | 25% |
| Rock | 22% |

**2.**

| Weekly Budget | |
|---|---|
| Meals | 22% |
| Savings | 17% |
| Music | 8% |
| Clothes | 35% |
| Movies | 12% |
| Books | 6% |

**3.**

| Daily Activities | |
|---|---|
| Sleeping | 9 hours |
| School | 7 hours |
| Eating | 2 hours |
| Recreation | 3 hours |
| Homework | 2 hours |
| Other | 1 hour |

**APPLICATIONS**   *Make a circle graph to show the data in each table.*

**4.**

| Area of Continents | |
|---|---|
| Africa | 20% |
| Antarctica | 9% |
| Asia | 30% |
| Australia | 6% |
| Europe | 7% |
| North America | 16% |
| South America | 12% |

**5.**

| Elements of Earth's Crust | |
|---|---|
| Oxygen | 47% |
| Silicon | 28% |
| Aluminum | 8% |
| Iron | 5% |
| Other | 12% |

**Name** _____  **Date** _____

# Using Statistics to Make Predictions

**W**hen real-life data are collected in a statistical experiment, the points graphed usually do not form a straight line. They may, however, approximate a linear relationship. A **best-fit line** can be used to show such a relationship. A best-fit line is a line that is very close to most of the data points.

**EXAMPLE**  *Use the best-fit line to predict the annual attendance at Fun Times Amusement Park in 2006.*

Draw a line so that the points are as close as possible to the line. Extend the line so that you can find the *y* value for an *x* value of 2006. The *y* value for 2006 is about 225,000.

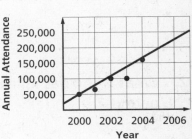

So, the annual attendance at Fun Times Amusement Park in 2006 is 225,000 people.

**Y**ou can also write an equation of a best-fit line.

**EXAMPLE**  *Use the information from the example above. Write an equation in slope-intercept form for the best-fit line and then predict the annual attendance in 2005.*

Step 1   First, select two points on the line and find the slope.
Choose (2001, 90,000) and (2003, 150,000).

$$m = \frac{y_2 - y_1}{x_2 - x_1}$$   *Definition of slope*

$$= \frac{150,000 - 90,000}{2003 - 2001}$$   $x_1 = 2001, y_1 = 90,000$
$x_2 = 2003, y_2 = 150,000$

$$= 30,000$$   *Simplify.*

Step 2   Find the *y*-intercept.
$$y = mx + b$$   *Slope-intercept form*
$$90,000 = 30,000(2001) + b$$   $y = 90,000, m = 30,000, x = 2001$
$$-59,940,000 = b$$   *Simplify.*

*(Continued on the next page.)*

Step 3    Write the equation.

$y = mx + b$                    *Slope-intercept form*

$y = 30,000x - 59,940,000$    $m = 30,000, b = -59,940,000$

Step 4    Solve the equation.

$y = 30,000(2005) - 59,940,000$        $x = 2005$

$= 210,000$                    *Simplify.*

The predicted annual attendance at Fun Times Amusement Park in 2005 is 210,000.

## EXERCISES

1. Predict the sales figure for 2006.

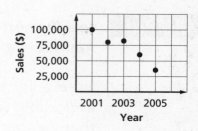

2. What is the gas mileage for a car weighing 4,500 pounds?

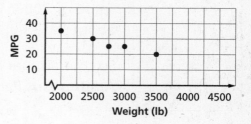

3. How tall would a tomato plant be ten days after planting the seed?

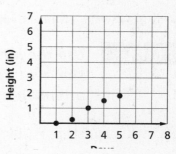

## APPLICATIONS

4. Use the graph in Exercise 2. Determine the equation of the best-fit line. Use it to predict the gas mileage for a car weighing 4,000 pounds.

5. Use the graph in Exercise 3. Determine the equation of the best-fit line. Use it to predict the height of the tomato plant 8 days after planting.

## SKILL 75

**Name** _____ **Date** _____

# Counting Outcomes and Tree Diagrams

**A**n organized list can help you determine the number of possible outcomes for a situation. One type of organized list is a **tree diagram.**

**EXAMPLE**  *The lunch special at Morgan's Diner is a choice of a turkey, ham, or veggie sandwich, salad, or fruit plate, and either pie or cake. If you wish to order the lunch special, how many different choices do you have?*

To answer this question, make a tree diagram.

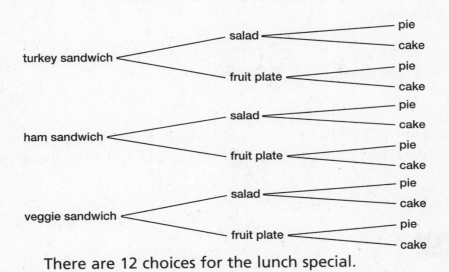

There are 12 choices for the lunch special.

**T**he **Fundamental Counting Principle** states that if an event M can occur *m* ways and it is followed by an event N that can occur *n* ways, then the event M followed by event N can occur *m* × *n* ways.

**EXAMPLE**  *If two quarters are tossed, find the total number of outcomes.*

Use the Fundamental Counting Principle.

There are 2 possible outcomes when tossing a coin, heads or tails.
outcomes for quarter 1 × outcomes for quarter 2 = possible outcomes
$$2 \quad\quad \times \quad\quad 2 \quad\quad = \quad\quad 4$$

There are 4 possible outcomes if 2 quarters are tossed.

*Draw a tree diagram to find the number of outcomes for each situation.*

1. A coin and a number cube are tossed.

2. The spinner is spun twice.

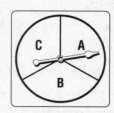

*Use the Fundamental Counting Principle to find the number of outcomes for each situation.*

3. 5 types of juice come in 3 different sized containers.

4. T-shirts come in 3 sizes and 6 colors.

5. Baseball hats come in 2 styles and 3 sizes for each of 12 teams.

6. Pizzas come in 5 sizes, with 2 different crusts and 14 available toppings.

## APPLICATIONS

7. Mrs. Jenkins' history test has 10 questions. Seven of the questions are multiple-choice with four answer choices. Two of the questions are true-false. How many possible sets of answers are there for the test?

8. Ryan is buying a new bicycle. He can choose from a mountain bike, a stunt bike, or a BMX bike. Each of the bikes comes in 6 colors. The bikes offer a choice of 2 types of tires and 3 types of seats. How many different bicycles can Ryan select?

9. Lonan is choosing a new password for his email account. The password must contain eight characters. The first two characters of his password must be letters and the last 6 digits must be any digit 0–9. How many possible passwords can Lonan create?

# Permutations

**A**n arrangement or listing in which order is important is called a **permutation**. $P(n, r)$ stands for the number of permutations of $n$ things taken $r$ at a time.

**EXAMPLE**

*There are 6 runners in a race. How many permutations of first, second, and third place are possible?*

There are 6 choices for first place, then 5 choices for second place, and finally 4 choices for third place. Find $P(6, 3)$

$$P(6, 3) = 6 \times 5 \times 4$$
$$= 120$$

The number of permutations is 120.

**T**he expression $6 \times 5 \times 4 \times 3 \times 2 \times 1$ can be written as 6!. It is read *six factorial*. In general, $n!$ is the product of whole numbers starting at $n$ and counting backward to 1. To find the number of permutations involving all members of a group, $P(n, n)$, find $n!$.

**EXAMPLE**

*There are 6 runners in a race.*
*In how many ways can they finish the race?*

There are 6 choices for first, 5 choices for second, and so on.

$$P(6, 6) = 6 \times 5 \times 4 \times 3 \times 2 \times 1 = 720$$

There are 720 ways in which the runners can finish the race.

**EXERCISES**  *Find the value of each expression.*

1.  $P(5, 2)$          2.  $P(8, 3)$          3.  $P(4, 3)$          4.  $P(7, 4)$

5.  $P(10, 2)$        6.  $P(4, 4)$          7.  $P(6, 1)$          8.  $P(9, 5)$

9.  $4!$                  10.  $5!$               11.  $8!$                12.  $10!$

13.  $\frac{6!}{3!}$          14.  $\frac{5!}{4!}$          15.  $\frac{6!}{2!}$          16.  $\frac{8!}{5!}$

17. How many ways can a winner and a runner-up be chosen from 8 show dogs at a dog show?

18. In how many ways can 5 horses in a race cross the finish line?

19. In how many ways can a president, vice-president, secretary, and treasurer be chosen from a club with 12 members?

20. A shelf has a history book, a novel, a biography, a dictionary, a cookbook, and a home-repair book. In how many ways can 4 of these books be rearranged on another shelf?

21. In how many ways can 8 people be seated at a counter that has 8 stools in a row?

22. Eight trained parrots fly onto the stage but find there are only 5 perches. How many different ways can the parrots land on the perches if only one parrot is on each perch?

23. Seven students are running for class president. In how many different orders can the candidates make their campaign speeches?

24. In how many different ways can a coach name the first three batters in a nine-batter softball lineup?

25. How many different flags consisting of 4 different-colored vertical stripes can be made up from blue, green, red, black, and white?

26. In how many ways can the gold, silver, and bronze medals be awarded to 10 swimmers?

**Name** _____ **Date** _____

# Combinations

Arrangements or listings in which order is *not* important are called **combinations**.

---

**EXAMPLE**

*A mail order catalog offers striped rugby shirts. Customers can choose two colors for the alternating stripes from a choice of red, blue, yellow, and white. How many ways can a customer order a rugby shirt?*

Since order is not important, this arrangement is a combination. List all the permutations of red, blue, yellow, and white taken two at a time. Then cross off arrangements that are the same as another one. For example, red and blue stripes are the same as blue and red stripes.

RB   RY   RW   B̶R̶   BY   BW
Y̶R̶   Y̶B̶   YW   W̶R̶   W̶B̶   W̶Y̶

There are only six different arrangements. So, there are six different ways a customer can order a rugby shirt.

---

$C(n, r)$ stands for the number of combinations of $n$ things taken $r$ at a time.

---

**EXAMPLE**

*In how many ways can 3 representatives be chosen from a group of 11 people?*

To find the number of combinations of choosing 3 representatives out of 11 people, first find the number of permutations of 11 things taken 3 at a time. Then divide this number by the number of ways 3 things can be arranged. You can use the formula $C(n, r) = \frac{P(n, r)}{r!}$.

$$C(11, 3) = \frac{P(11, 3)}{3!}$$

$$= \frac{11 \times 10 \times 9}{3 \times 2 \times 1} \qquad \textit{Find P(11, 3).}$$

$$= \frac{990}{6} \qquad \textit{Divide by 3! to eliminate combinations that are the same except for order.}$$

$$= 165$$

---

*Find the value of each expression.*

1. $C(10, 9)$      2. $C(8, 8)$      3. $C(4, 2)$      4. $C(5, 3)$

5. $C(6, 2)$      6. $C(6, 4)$      7. $C(7, 3)$      8. $C(8, 6)$

9. $C(12, 8)$      10. $C(15, 6)$      11. $C(10, 4)$      12. $C(9, 3)$

**APPLICATIONS**

13. For an English exam, students are asked to write 4 essays from a list of 8 topics. How many different combinations are possible?

14. In how many ways can a starting team of 5 players be chosen from a team of 14 players?

15. The Supreme Court has 9 justices. At least 5 of the justices must agree on a decision.
    a. How many different combinations of 5 justices can agree on a decision?

    b. How many different combinations of 6 justices can agree on a decision?

    c. How many different combinations of 7 justices can agree on a decision?

    d. How many different combinations of 8 justices can agree on a decision?

16. In the Illinois State Lottery, balls are numbered from 1 to 54 and put into a machine, which scrambles them. Six balls are then selected in any order. How many different ways can the winning number be chosen?

17. How many different five-card hands are possible using a standard 52-card deck?

# Probability

The **probability** of an event is the ratio of the number of ways an event can occur to the number of possible outcomes. The probability of one event occurring is called a **simple probability**.

---

**EXAMPLE**

*The spinner has ten equally likely outcomes. Find the probability of spinning a number less than 7.*

Numbers less than 7 are 1, 2, 3, 4, 5, and 6.
There are 10 possible outcomes.

probability of a number less than 7 = $\frac{6}{10}$ or $\frac{3}{5}$

The probability of spinning a number less than 7 is $\frac{3}{5}$.

---

A **compound event** consists of two or more simple events. **Independent events** occur when the outcome of one event does *not* affect the outcome of another event. If the outcome of one event affects the outcome of another event, the events are called **dependent events**.

---

**EXAMPLES**

*Tiles numbered 1 through 25 are placed in a box. Two tiles are selected at random. Find each probability.*

*drawing an even number, replacing the tile, and then randomly drawing a multiple of 3*

The events are independent since the outcome of one drawing does not affect the other.

$P(\text{even number}) = \frac{12}{25}$ $\qquad$ $P(\text{multiple of 3}) = \frac{8}{25}$

$P(\text{even number, multiple of 3}) = \frac{12}{25} \cdot \frac{8}{25}$ or $\frac{96}{625}$

*drawing a number greater than 10, and then drawing a number less than 10 without replacing the first tile*

The events are dependent since there is one less tile from which to choose on the second draw.

$P(n > 10) = \frac{15}{25}$ or $\frac{3}{5}$ $\qquad$ $P(n < 10) = \frac{9}{24}$ or $\frac{3}{8}$

$P(n > 10, n < 10) = \frac{3}{5} \cdot \frac{3}{8}$ or $\frac{9}{40}$

---

*The spinner shown is equally likely to stop on each of the sections. Find each probability.*

1.  $P(n < 5)$

2.  $P$(even number)

3.  $P$(multiple of 4)

4.  $P$(factor of 6)

*A bag of marbles contains 3 yellow, 6 blue, 1 green, 12 red, and 8 orange marbles. Find each probability.*

5.  $P$(red)

6.  $P$(blue or yellow)

7.  $P(not$ orange)

8.  $P$(yellow then blue, with replacement)

9.  $P$(green then red without replacement)

10.  $P$(yellow, yellow, orange, without replacement)

*The spinner shown is equally likely to stop on each of the sections. The spinner is spun twice. Find each probability.*

11.  $P$(multiple of 2, multiple of 3)

12.  $P(n > 10, n > 12)$

13.  $P$(product is even)

14.  $P$(sum = 20)

*A number cube is rolled and the spinner is spun. Find each probability.*

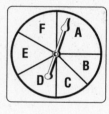

15.  $P$(6 and B)

16.  $P$(odd number and E)

17.  $P(n > 3$ and A, B, or C)

18.  $P$(n < 3 and vowel)

*Brianna, Mai-Lin, and Camila are playing a board game in which two number cubes are tossed to determine how far a player's game piece is to move.*

19.  Brianna needs to move her piece 9 spaces to return it to base. What is the probability that she will roll 9?

20.  If Mai-Lin rolls doubles, then she gets to roll again. What is the probability that Mai-Lin will get to roll twice on her next turn?

**Name** _____ **Date** _____

# Theoretical and Experimental Probability

The **theoretical probability** of an event is the ratio of the number of ways the event can occur to the number of possible outcomes.

The **experimental probability** of an event is the ratio of the number of successful trials to the number of trials.

---

**EXAMPLE**

*Sean wants to determine the probability of getting a sum of 7 when rolling two number cubes. The sample space, or all possible outcomes, for rolling two number cubes is shown below.*

```
1, 1   1, 2   1, 3   1, 4   1, 5   1, 6
2, 1   2, 2   2, 3   2, 4   2, 5   2, 6
3, 1   3, 2   3, 3   3, 4   3, 5   3, 6
4, 1   4, 2   4, 3   4, 4   4, 5   4, 6
5, 1   5, 2   5, 3   5, 4   5, 5   5, 6
6, 1   6, 2   6, 3   6, 4   6, 5   6, 6
```

*What is the theoretical probability of rolling a sum of 7? What is the experimental probability of rolling a sum of 7 if Sean rolls the number cubes 20 times and records 4 sums of 7?*

There are 6 sums of 7 shown in the sample space above. So, the theoretical probability of rolling a sum of 7 is $\frac{6}{36}$ or $\frac{1}{6}$.

Since Sean rolled 4 sums of 7 on 20 rolls, the experimental probability is $\frac{4}{20}$ or $\frac{1}{5}$.

---

**EXERCISES** *Find the theoretical probability of each of the following.*

1. getting tails if you toss a coin

2. getting a 6 if you roll a number cube

3. getting a sum of 2 if you roll two number cubes

---

**4.** getting a sum less than 6 if you roll two number cubes

**5.** Amanda rolled one number cube 30 times and got 8 sixes.

    **a.** What is her experimental probability of getting a six?

    **b.** What is her experimental probability of *not* getting a six?

**6.** Ramón rolled two number cubes 36 times and got 3 sums of 11.

    **a.** What is his experimental probability of getting a sum of 11?

    **b.** What is his experimental probability of *not* getting a sum of 11?

**APPLICATIONS**   *While playing a board game, Akira rolled a pair of number cubes 48 times and got doubles 10 times.*

**7.** What was his experimental probability of rolling doubles?

**8.** How does his experimental probability compare to the theoretical probability of rolling doubles?

**9.** How do you think the experimental probability compares to the theoretical probability in most experiments?

**10.** Do you think the experimental probability is ever equal to the theoretical probability? Explain?

Name _____ Date _____

# Problem-Solving Strategies

Three possible strategies for solving problems are listed below.
• Guess and Check     • Look for a Pattern     • Eliminate the Possibilities

**EXAMPLE**

*The Connor family bought some tickets to the zoo. Admission is $12 for adults and $7 for children under 12. They spent $71 for admission. How many adult tickets and how many children's tickets did the Connor family buy?*

Use the guess-and-check strategy to solve the problem. Suppose your first guess is 2 adults and 5 children.

$$2 \times 12 + 5 \times 7 = 59$$

This guess is too low. Try 3 adults and 6 children.

$$3 \times 12 + 6 \times 7 = 78$$

This guess is too high. Try 3 adults and 5 children.

$$3 \times 12 + 5 \times 7 = 71$$

The Connor family bought 3 adult tickets and 5 children tickets.

**EXAMPLE**

*Kwan gave the clerk $60 to pay for a purse that costs $32.95 and a hat that costs $17.50. Should she expect about $10, $20, or $30 in change?*

In this problem, an exact answer is not needed. Use the eliminate-the-possibilities strategy.

First estimate the answer by rounding $32.95 to $33 and $17.50 to $18. Kwan spent about $33 + $18 or $51. Since $60 − $51 = $9, you can eliminate $20 and $30 as possible answers.

The correct answer is about $10.

## EXERCISES   *Solve.*

1. Fill in the boxes at the right with the digits 0, 1, 3, 4, 5, and 7 to make a correct multiplication problem. Use each digit exactly once.

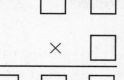

2. Write the next two numbers in the pattern.

   $\frac{1}{2}, \frac{2}{3}, \frac{3}{4}, \frac{4}{5} \cdots$

3. Does $81.4 \times 0.68$ equal 553.52, 55.352, or 5.5352? Do *not* actually compute.

4. What is the total number of rectangles in the figure at the right?

## APPLICATIONS

5. Abby's test scores were 95, 82, 78, 84, and 88. Is the best estimate of her average test score 90, 85, 70, or 75?

6. Erica has some quarters and dimes in her pocket. The value of the coins is $1.65. If she has a total of 9 coins, how many quarters and how many dimes does Erica have?

7. James wants to work up to doing 40 sit-ups a day. He plans to do 5 sit-ups the first day, 9 sit-ups the second day, 13 sit-ups the third day, and so on. On what day will he do 45 sit-ups?

8. The school bell rings at 8:05 A.M., 8:47 A.M., 8:50 A.M., 9:32 A.M., 9:35 A.M., 10:17 A.M., 10:20 A.M., and 11:02 A.M. If the pattern continues, what are the next three times the bell will ring?

9. Armando bought a car. He paid $3,000 down and will pay $350 per month for 48 months. Does the car cost closer to $30,000, $25,000, $20,000, or $15,000?

10. Jennifer bought some cookies for 55¢ each and some bottles of fruit drink for 80¢ each. She spent $5.70. How many cookies and bottles of fruit drink did she buy?

11. The length of a rectangle is 6 more inches than the width. The area of the rectangle is 216 square inches. What are the dimensions of the rectangle?

## SKILL 81

**Name** _____ **Date** _____

# Work Backward

$S$ome problems start with the end result and ask for something that happened earlier. The strategy of **working backward**, or **backtracking**, can be used to solve problems like this. To use this strategy, start with the end result and undo each step.

**EXAMPLE**  *A number is decreased by 12. The result is multiplied by 5, and 30 is added to the new result. The final result is 200. What is the number?*

Use a flowchart to show the steps in the computation.

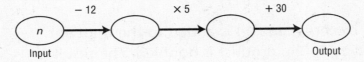

Find the solution by starting with the output.

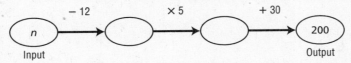

Since 30 was added to get 200, subtract 30. $200 - 30 = 170$

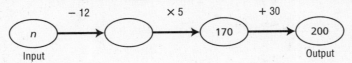

Next, divide 170 by 5. $170 \div 5 = 34$

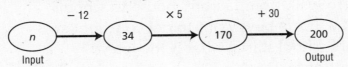

Then, add 12 to 34. $34 + 12 = 46$

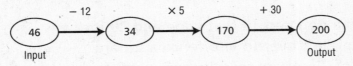

Thus, the number is 46.

## EXERCISES  *Solve by working backward.*

1. A number is added to 12, and the result is multiplied by 6. The final answer is 114. Find the number.

2. A number is divided by 3, and the result is added to 20. The result is 44. What is the number?

3. A number is divided by 8, and the result is added to 12. The final answer is 78. Find the number.

4. Twenty five is added to a number. The sum is multiplied by 4, and 35 is subtracted from the product. The result is 121. What is the number?

5. A number is divided by three, and 14 is added to the quotient. The sum is multiplied by 7. The product is doubled. The result is 252. What is the number?

## APPLICATIONS

6. A bacteria population doubles every 8 hours. If there are 1,600 bacteria after 2 days, how many bacteria were there at the beginning?

7. Each school day, Alexander takes 35 minutes to get ready for school. He takes 5 minutes to walk to Jaaron's house. The two boys take 15 minutes to walk from Jaaron's house to school. School starts at 8:10 A.M. If the boys want to get to school at least 10 minutes before school starts, what is the latest Alexander must get out of bed?

8. A fence is put around a dog pen 10 feet wide and 20 feet long. Enough fencing is left over to also fence a square garden with an area of 25 square feet. If there are 3 feet left after the fencing is completed, how much fencing was available at the beginning?

**Name** _____  **Date** _____

# Simplify and Use Logical Reasoning

Two possible strategies for solving problems are listed below.
- Solve a Simpler Problem
- Use Logical Reasoning

---

**EXAMPLE**  *Find the sum of the whole numbers from 1 to 200.*

This would be a tedious problem to solve using a calculator or adding the numbers yourself. The problem is easier to solve if you solve a simpler problem. First, consider the partial sums indicated below.

1, 2, 3, ..., 100, 101, ..., 198, 199, 200

$$100 + 101 = 201$$

$$3 + 198 = 201$$
$$2 + 199 = 201$$
$$1 + 200 = 201$$

Notice that each sum is 201. There are 100 of these partial sums.

$$201 \times 100 = 20{,}100$$

The sum of the whole numbers from 1 to 200 is 20,100.

---

**EXAMPLE**  *Neva, Justin, Toshiro, and Sydney are lining up by height. Toshiro is not standing next to Sydney. Neva is the shortest and is not standing next to Toshiro. List the students from shortest to tallest.*

Use logical reasoning to solve this problem.
- Since Neva is the shortest, write Neva in the first position.
- Since Toshiro and Sydney are not standing next to each other, write Justin in the third position.
- Since Neva and Toshiro are not standing next to each other, write Toshiro in the fourth position.
- Since Sydney is the only student left, write Sydney in the second position.

**1.** __Neva__  **2.** __Sydney__  **3.** __Justin__  **4.** __Toshiro__

The students from shortest to tallest are Neva, Sydney, Justin, and Toshiro.

---

## EXERCISES   *Solve.*

1. Find the sum of the whole numbers from 1 to 400.

2. Find the sum of the even numbers from 2 to 100.

3. There are three boards each a different odd number of feet long. If the boards are placed end to end, the total length is 9 feet. What are the lengths of the boards?

4. A total of 492 digits are used to print all the page numbers of a book beginning with page 1. How many pages are in the book?

5. Anna, Iris, and Oki each have a pet. The pets are a fish, cat, and a bird. Anna is allergic to cats. Oki's pet has 2 legs. Whose pet is the fish?

## APPLICATIONS

6. Connie, Kristina, and Roberta are the pitcher, catcher, and shortstop for a softball team, but not necessarily in that order. Kristina is not the catcher. Roberta and Kristina share a locker with the shortstop. Who plays each position?

7. Mr. Lee wants to carpet the room shown at the right. How much carpet will he need?

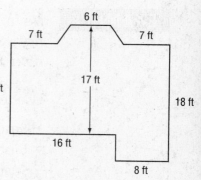

8. Doug, Louann, and Sandy have lockers next to each other. Louann rides the bus with the person whose locker is at the right. Doug's locker is not next to Luann's locker. Who has the locker at the left?

9. A rectangular field is fenced on two adjacent sides by a brick wall. The field is 63 yards long with an area of 1,323 square yards. How much fencing is needed on the two sides of the field?

10. A vending machine sells items that cost 80¢. It only accepts quarters, dimes, and nickels. If it only accepts exact change, how many different combinations of coins must the machine be programmed to accept?

Name _____  Date _____

# Organizing Information

Two possible ways to organize information to solve problems are listed below.
- Make a List                    • Use a Matrix or Table

**EXAMPLE**

*On Saturday, Omar plans to go to the library, the discount store, and his grandmother's house. He cannot decide in which order to go to these locations. How many choices does he have?*

Make a list to show the different orders.

library, discount store, grandmother's house
library, grandmother's house, discount store
discount store, library, grandmother's house
discount store, grandmother's house, library
grandmother's house, library, discount store
grandmother's house, discount store, library

Omar has 6 choices for the order he can go to the locations.

**EXAMPLE**

*Kimi is offered two jobs. Job A has a starting salary of $24,000 per year with a guaranteed raise of $1,200 per year. Job B has a starting salary of $26,000 with a guaranteed raise of $800 per year. In how many years will both jobs pay the same amount of money?*

Make a table to show the effect of each option over a 6-year period.

| Year | Job A | Job B |
|------|--------|--------|
| 1 | $24,000 | $26,000 |
| 2 | $25,200 | $26,800 |
| 3 | $26,400 | $27,600 |
| 4 | $27,600 | $28,400 |
| 5 | $28,800 | $29,200 |
| 6 | $30,000 | $30,000 |

In 6 years, the two jobs will pay the same amount of money.

1. How many different four-digit numbers can be formed using each of the digits 1, 2, 3, and 4 once?

2. How many different two-digit numbers can be formed using 1, 2, 3, 4, and 5 if the digits can be used more than once?

3. How many ways can you give a clerk 65¢ using quarters, dimes, and/or nickels?

## APPLICATIONS

4. Rebecca, Lisa, and Courtney each have one pet. The pets are a dog, a cat, and a parrot. Courtney is allergic to cats. Rebecca's pet has two legs. Whose pet is the dog?

5. Mr. Ramos is starting a college fund for his daughter. He starts out with $700. Each month he adds $80 to the fund. How much money will he have in a year?

6. The Centerville Civic Association is selling pizzas. They can add pepperoni, green peppers, and/or mushrooms to their basic cheese pizzas. How many different kinds of pizzas can they sell?

7. Kyle, Gabrielle, Spencer, and Stephanie each play a sport. The sports are basketball, gymnastics, soccer, and tennis. No one's sport starts with the same letter as his or her name. Gabrielle and the soccer player live next door to each other. Stephanie practices on the balance beam each day. Gabrielle does not own a racket. Which student plays each sport?

8. A deli sells 5 different soft drinks in 3 different sizes. How many options does a customer have to buy a soft drink?

9. In the World Series, two teams play each other until one team wins 4 games.
   a. What is the greatest number of games needed to determine a winner?

   b. What is the least number of games needed to determine a winner?

**Name** _____    **Date** _____

# Visualizing Information

Three possible ways to visualize information to solve problems are listed below.
- Draw a Diagram
- Use a Graph
- Make a Model

**EXAMPLE**

*On the first day, Derrick e-mailed a joke to 3 of his friends. On the second day, each of these friends e-mailed 3 other people. On the third day, each of the people who read the joke on the second day e-mailed 3 more people. By the end of the third day, how many people read the joke?*

To solve the problem, draw a diagram.

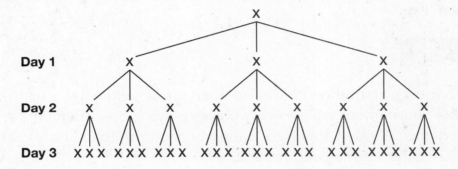

Count the number of Xs in Day 1, Day 2, and Day 3. By the end of the third day 39 people read the joke. Note that the first person e-mailed the joke, but did not read it, during the three days.

**EXAMPLE**

*Identical boxes are stacked in the corner of a room as shown below. How many boxes are there altogether?*

Make a model using cubes and count the number of cubes.

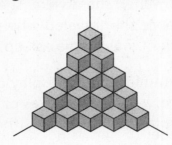

If you modeled the problem correctly, there should be 35 boxes.

1. How many different shapes of rectangular prisms can be formed using exactly 18 cubes?

2. Six points are marked around a circle. How many straight lines must you draw to connect every point with every other point?

3. A cube with edges 4 inches long is painted on all six sides. Then, the cube is cut into smaller cubes with edges 1 inch long as shown at the right.

   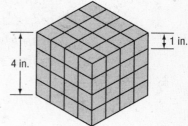

   a. How many of the smaller cubes are painted on only one side?

   b. How many of the smaller cubes are painted on exactly two sides?

   c. How many of the smaller cubes have no sides painted?

## APPLICATIONS

4. Coach Robinson is the tennis coach. He wants to schedule a round-robin tournament where every player plays every other player in singles tennis. If there are 8 members on the team, how many matches should the coach schedule?

5. The graph shows the population growth of Anchorage, Alaska.

   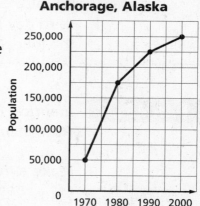

   **Population of Anchorage, Alaska**

   a. During what 10-year period did Anchorage show the greatest growth in population?

   b. What would you estimate the population will be in 2010?

6. Halfway through her plane flight from New York City to Orlando, Emma fell asleep. When she awoke, she still had to travel half the distance she traveled when asleep. For what fraction of the flight was Emma asleep?

7. Emilio wants to make a pyramid-shaped display of soccer balls for his sporting goods store. How many boxes of soccer balls will he need to make a display like the one at the right?